The covenant of reason

Isaac Levi is one of the preeminent philosophers in the areas of pragmatic rationality and epistemology. This collection of essays constitutes an important presentation of his original and influential ideas about rational choice and belief. A wide range of topics is covered, including consequentialism and sequential choice, consensus, voluntarism of belief, and the tolerance of the opinions of others.

The essays elaborate on the idea that principles of rationality are norms that regulate the coherence of our beliefs and values with our rational choices. The norms impose minimal constraints on deliberation and inquiry, but they also impose demands well beyond the capacities of deliberating agents. The first group of essays reflects on the gulf between the demands of reason and the incapacity of agents to fulfill those demands, and on the role of commitment in this gulf. A corollary of this view is that deliberation crowds out prediction, and a second group of essays explains the ramifications of this claim for theories of sequential choice, diachronic rationality, and notions of equilibrium in game theory. A third group of essays examines the condition of doubt regarding judgments of truth, probability and value, and the parallel notion of consensus amid disagreement. The final two essays gather all these themes together in exploring value pluralism and different ways of responding to dissent.

This major collection will be eagerly sought out by a wide range of philosophers in epistemology, logic, ethics, and philosophy of science, as well as economists, decision theorists, and statisticians.

The covenant of reason

Rationality and the commitments of thought

ISAAC LEVI
Columbia University

PUBLISHED BY THE PRESS SYNDICATE OF THE UNIVERSITY OF CAMBRIDGE
The Pitt Building, Trumpington Street, Cambridge CB2 1RP, United Kingdom

CAMBRIDGE UNIVERSITY PRESS
The Edinburgh Building, Cambridge CB2 2RU, United Kingdom
40 West 20th Street, New York, NY 10011-4211, USA
10 Stamford Road, Oakleigh, Melbourne 3166, Australia

First published 1997

Printed in the United States of America

Typeset in Garamond Book

Library of Congress Cataloging-in-Publication Data

Levi, Isaac, 1930-

The covenant of reason : rationality and the commitments of thought/Isaac Levi.

p. cm.

Includes bibliographical references.

ISBN 0-521-57288-6 (hardcover). - ISBN 0-521-57601-6 (pbk.)

1. Reasoning. 2. Reason. 3. Norm (Philosophy) 4. Rational choice theory. 5. Belief and doubt. I. Title.
BC177.L48 1997
128'.33 - dc21 97-2958
CIP

A catalog record for this book is available from the British Library.

ISBN 0 521 57288 6 hardback
ISBN 0 521 57601 6 paperback

Contents

Introduction

Rationality, according to some, is an excess of reasonableness. We should be rational enough to confront the problems of life; but there is no need to go whole hog. Indeed, doing so is something of a vice.

Obsessive behavior that insists on computing to the 100th decimal place when the demands of the situation call for determining no more than 2, may, indeed, be an obstacle to effective agency. We need not explicitly or consciously seek to identify the implications of our beliefs and values any further than is required in order to address the problems we currently anticipate facing. But defenders of rationality do not say otherwise when they claim that rational agents ought to recognize the implications of their beliefs and values. Obviously, this obligation applies "when and to the extent that the need arises." However, there is no upper bound on how complex the problems are that we may need to face. As a consequence, it is always desirable to improve our capacities to confront ever more sophisticated tasks. Advocates of reasonableness seem to deny or overlook this point.

Such denial would make sense if we led lives so simple that we could get by without taxing our ratiocinative capacities. There would then be no need to improve our capacities to reason well, for the capacities of human agents would never be stretched to the limit by the complexities of life and nature. But there always will be such a need. The advocates of reasonableness as a replacement for rationality would appear to be out of touch with reality.

Others acknowledge and, indeed, emphasize the lack of an upper bound on the complexity of the problems we may face while emphasizing "boundedness" of our rationality. They suggest that we make a virtue out of necessity by replacing the demands of rationality with simpler requirements that are within the grasp of ordinary people and still allow us to cope.

This maneuver is self-defeating. No matter what nontrivial standards for bounded rationality are endorsed, problems can arise in real life that cannot be solved in a manner satisfying the standards with the computational power available to us.

The partisans of reasonableness and of bounded rationality share much in common. The use of principles of logic, rational choice and probability judgment as characterizations of demands of coherence on beliefs, probability judgments and values is dismissed in one way or another as of marginal relevance to reflective thought. Such principles are neither explanatory nor predictive of human behavior. When these principles are construed as a system of prescriptions regulating our beliefs, values and choices, they lose touch with human capacity. Principles of logic, rational choice and probability judgment may do very well as systems of truths about some domain or other but they have neither descriptive nor prescriptive relevance to human behavior.

The essays in this volume proceed from a perspective opposed to these conclusions. Logic and principles of rationality receive their philosophically most important application in regulating how agents ought to think. Rationality urges us to be aware of what we believe and the implications of such beliefs, and it urges us to do so consistently.

Thus, someone who believes that New York has a larger population than Los Angeles has undertaken to believe that New York has a larger population than Detroit, if the agent believes that Los Angeles is larger than Detroit and that "has a larger population than" is transitive. If the agent has made the undertaking, the agent has incurred an obligation to fulfill the commitment.

The demands of rationality on probability and value judgment differ in technical detail; but take on a shape similar to the demands on rational belief.

The advocates of reasonableness and boundedness of rationality are right to point to our limited capacities. Instead of despairing as they do over the prospects for improvement, we ought to recognize our infirmities and seek out therapies, training regimens, and technologies that can extend our capacities to achieve coherence and closure. Rational agents, like religious agents, undertake to satisfy standards that transcend their capacities to realize. Failing to realize the standards is no sin. Failing to try harder is.

Much of my previous writing has addressed issues concerning the conditions under which *changes* in belief and value are justified. These changes were explicitly understood to be *undertakings* – that is, changes in *commitments* to judgments of truth and value. Changes in *performances fulfilling commitments already undertaken* were not considered except in passing.

Chapter 2 on *The Fixation of Belief and Its Undoing* breaks with this emphasis on commitment to the exclusion of performance. I argued there that empirical investigations of changes in believing and valuing that are changes in performance without being changes in commitment are of vital importance for clinical and technological reasons. The more accurately we understand the sources of our failures to be rational (i.e., to fulfill our doxastic and other attitudinal commitments) and the resources at our disposal for overcoming these limitations, the more accurate will be our appreciation of when it is and is not quixotic to undertake efforts to develop ways and means to enhance our capacity to fulfill our commitments and thereby to be rationally coherent.

Most of the essays in this volume elaborate on topics related to being rational in the sense of coherently or consistently fulfilling the commitments generated by our acts and words according to principles of rationality.

According to the thesis presented in Chapter 1, principles of rationality characterize the propositional attitudes by specifying the terms of the contract made by the agent who adopts an attitudinal commitment in the context of other beliefs and values.

Chapter 2 emphasizes an important feature of rational agency. Rational agents are responsible in deliberation for policing the coherence of their conclusions. Insofar as a self-critical agent invokes principles of rationality in rational deliberation to bring his or her own reflections under critical control and to determine which of the options available in a given setting are admissible relative to the agent's beliefs and values, these principles cannot also be used by the agent to predict and explain the agent's choices. Deliberation crowds out prediction.

A corollary of the principal claims of Chapters 1 and 2 is that principles of rationality are neither predictive nor explanatory laws. To be sure, once one assumes that someone is conforming to the requirements of rationality in a given setting, one may sometimes predict or explain his or her behavior. But explaining why someone's beliefs and goals influenced his or her actions in a certain way by noting that the agent was rational at the moment is tantamount to explaining why taking opium induced sleep by citing the dormitive virtue of opium.

In Chapter 3, fulfilling the commitments undertaken when propositional attitudes are endorsed requires perfect consistency or coherence in these attitudes. Principles of rationality are taken to be prescriptions concerning how one ought to think – i.e., principles of logic in the "qualified psychologistic" sense that Carnap thought was a trivial consequence of an objectivist conception of logic.

Carnap's view reflects Frege's opinion that the "logic of consistency," as Ramsey understood it, coincides with "the logic of truth." Ramsey challenged Keynes's attempt to introduce a probability logic that would mimic deductive logic in this way. In Chapter 3, I argue that the logic of consistency for full belief or certainty has the structure of an S5 modal logic (for which the alethic condition in S5 holds) that does not coincide with a logic of truth for full belief (for which the alethic condition fails). Insofar as even classical deductive logic prescribes how agents ought to think, it would appear to do so as a fragment of a logic of full belief. Insofar as students of doxastic and epistemic logic have embraced visions of validity that presuppose the coincidence of a logic of consistency and a logic of truth, they are guilty of a serious confusion.

The claim that deliberation crowds out prediction of one's own choices and vice versa has ramifications for theories of sequential choice where it is often taken for granted that the long-term plan for making a series of choices that is to be recommended in a decision problem in "extensive form" should coincide with the corresponding single-shot "up front" option in the associated "normal form" decision problem. The economist Peter Hammond has shown that if rational agents ought to equate extensive and normal form, they are committed to having complete orderings of their options. Hammond claims that students of rational choice understand consequentialism to entail equivalence of extensive and normal form. Consequentialism, therefore, demands that conditions of choice consistency and weak ordering be satisfied. Since the account of decision making based on indeterminate probabilities and utilities I have advocated since the early 1970s allows for noncomparability in the values of options and violation of choice consistency conditions, Hammond's conclusion implies that my view is incompatible with consequentialism.

Chapter 4 argues that because deliberation crowds out prediction of one's own choice, the extensive-normal form equivalence assumption is untenable. This result implies that Hammond's version of consequentialism ought to be rejected. However, there are other versions of consequentialism whose status is untouched by the critique of Hammond's consequentialism and some of them are identified.

Undermining the extensive-normal form equivalence not only allows for recognition of the coherence of indeterminate probability and utility judgment but also strips dynamic Dutch book arguments seeking to support applications of Bayes' theorem in the updating of probabilities via conditionalization of their obligatory character as conditions of "dynamic coherence". In general, I think that coherence or consistency is a "synchronic" property of beliefs or other attitudes endorsed by an agent at a given time and not a "diachronic" property of a sequence of attitudes that have been changed over time. The relevance of rationality to the question of change of point of view is that when deliberating as to what to do or as to how to change one's mind, an agent's judgments as to what he should do or how he or she should change endorsed at t_0 prior to decision and implementation are constrained by the beliefs and values endorsed by the agent at t_0 according to principles of synchronic rationality. There are no principles of diachronic rationality specifying conditions of consistency for beliefs, values and decisions over time.

The argument showing that deliberation crowds out prediction not only undermines the extensive-normal form equivalence but implies the incoherence of the decision maker's predicting that he will choose rationally either with certainty or with probability.

Chapter 5 argues that the same doubts about predicting one's own choice in the context of deliberation undermines several of the premises of Aumann's argument from his correlated equilibrium theorem to the conclusion that Bayes's rational players in a game must attain a correlated equilibrium.

Chapter 6 is a reissue of my first paper on indeterminate probability. I interpret indeterminacy in probability and value judgment as doubt or suspense with respect to probability or value. The position taken here disagrees with Ramsey's thought that the logic of consistency for probability (and value) judgment has no room for such indeterminacy.

Chapter 7 points to the analogy between suspense with respect to probability judgment and consensus between agents with respect to probability judgment. It argues that if two or more agents who differ in their probability judgments at t_0 seek to resolve their differences, they should do so through inquiry starting from a consensus as shared agreements at the outset of inquiry and terminating with a consensus reached at the end of inquiry. This model is contrasted with a proposal of Lehrer and Wagner. An objection raised by Laddaga showing that Lehrer and Wagner cannot handle the problem of relevance very well is discussed. It is shown that representing doubt regarding probability judgments by convex sets of probability functions poses no obstacle to representing irrelevance.

Chapters 8 and 9 elaborate further on the notions of consensus in probability and in utility judgment. Chapter 8 argues from the demand that when two or more agents evaluate the expected utilities of options differently due to disagreements in probability (utility) judgment against a background of agreement concerning utility (probability) judgment and seek to identify a consensus, that consensus should be representable by the set of expected utility functions over all mixtures of the commonly available options that preserve the preferences among options in the mixture set according to Pareto unanimity. The set of such expectation functions over options and the set of utility functions over the consequences should both be convex.

Seidenfeld, Kadane and Schervish subsequently pointed out that when parties to a joint decision problem differ with respect to both probability judgments over states and utilities for outcomes, convexity cannot be derived from Bayesian requirements; such requirements stipulate that potential resolutions of the differences in valuations of options satisfy the coherence requirements advanced by advocates of the prescription that expected utility be maximized while preserving the shared agreements between the parties understood as preserving Pareto unanimity. To save convexity, one needs to modify Pareto unanimity in ways that Seidenfeld, Kadane and Schervish find to be of doubtful merit. Chapter 9 argues that one needs to take into account more than Pareto unanimity in order to obtain an acceptable account of consensus.

Chapter 10 applies the proposals I have made concerning decision making under unresolved conflict to situations where experimental subjects appear to violate the "independence postulate" of expected utility theory. The apparent violation emerges when the subjects' choices are taken to reveal their preferences. I make it plain in Chapters 4 and 6 that I do not think that satisfying standard requirements of "choice consistency" is the hallmark of rationality; but I also insist that an agent's attitudes of full belief, probability judgment and value judgment do meet standards of rational coherence including requirements that preferences over lotteries satisfy the independence postulate. The immediate consequence is that value or preference is not to be confused with revealed preference. The criteria for choice I have considered do indicate ways to link preference including indeterminate preference with choice and revealed preference without conflating the two ideas.

Chapters 6–10 emphasize that consensus and doubt with respect to probability and value are not to be condemned as irrational or incoherent because they preclude weak ordering of options in decision problems. The idea that rational agents choose for the best all things considered is a dogma shared by economists, decision theorists and moral philosophers of otherwise diverse persuasions.

My resistance to this dogma derives in part from worries similar to those registered by Charles Peirce in objecting to the "Bayesians" of his day. It is also influenced by my admiration for John Dewey's attempt to extend the belief-doubt model of inquiry pioneered by Peirce for scientific inquiry to cover inquiry into values. Dewey insisted that there is no single standard of value as utilitarians and Kantians alike would insist. Because of the diversity of value, agents are bound to find themselves in conflicts that will occasion moral reflection or inquiry.

Agents confronting such difficulties may have to make decisions concerning how to act prior to reaching a satisfactory resolution of such conflicts through inquiry. Instead of supposing that conflicts are resolved by the choices made, I argue that truly rational agents ought to admit that the choices they make are not made for the best. Relative to the information available to them, there is no best.

Chapter 11 comments on the relation between the value pluralism of John Dewey and the value pluralism of Isaiah Berlin and Bernard Williams. Both pluralisms recognize that the plurality of values opens up the possibility of conflicts in value. For Dewey, such conflict in value is an occasion for doubt and inquiry. Decisions may have to be made without a satisfactory termination of inquiry, but that does not warrant failure to recognize the importance of finding satisfactory resolutions. The value pluralism of Isaiah Berlin and Bernard Williams fails to take moral inquiry seriously.

Chapter 12 closes this volume with a discussion of dissent. Tolerating dissent while failing to take it seriously is contrasted with respecting dissent by inquiring into the issues raised by the dissenter. Liberalism need not insist that all dissent should be taken seriously. There has to be a good reason to do so. I suggest that good reasons for taking dissent seriously parallel the sorts of good reasons that justify someone ceasing to take for granted what he or she initially regarded as certainly true and free of serious doubt. Certification of the authority of the dissenter furnishes a good reason for opening up one's mind to his view even if it does not justify slavish deference to it. And sometimes one can recognize a dissenter's view to be a bright idea meriting closer scrutiny even if the dissenter's credentials are of obscure provenance.

I included Chapters 11 and 12 in this collection of essays on synchronic rationality, coherence and logicality for two reasons: These two chapters discuss in one way or another changes in commitment instituted through inquiry. In addition, both chapters serve to emphasize that questions of identifying ideals of well-conducted inquiry are not to be relegated to methodologists of science. They have a bearing on how we are to understand moral conflict and political dissent.

All but two of the essays printed here have been published elsewhere. The provenances of all of these essays are given here:

- "Rationality and Commitment" (Chapter 1) is reprinted with kind permission from *Artifacts, Representations and Social Practice, BSPS,* vol. 154, edited by C. C. Gould and R. S. Cohen, Dordrecht: Kluwer (1994), 257-75.
- "Rationality, Prediction and Autonomous Choice" (Chapter 2) is reprinted with kind permission from *Canadian Journal of Philosophy,* suppl. vol. 19 (1989), 339-63.
- "The Logic of Full Belief" (Chapter 3) is to appear as a contribution to a collection of essays in honor of Charles Parsons, edited by G. Sher and R. Tieszen. I am grateful to the editors for granting me permission to publish it in this collection as well.
- "Consequentialism and Sequential Choice" (Chapter 4) is an extended and modified version of an essay that appeared in *Foundations of Decision Theory,* edited by M. Bacharach and S. Hurley, Blackwell (1991), 92-122.
- "Prediction, Deliberation and Correlated Equilibrium" (Chapter 5) is a version of a paper that was delivered in Vienna in July 1996 at a meeting sponsored by the Wiener Kreis Institut on Rationality.
- "On Indeterminate Probabilities" (Chapter 6) first appeared in *The Journal of Philosophy* 71 (1974), 391-418. The version appearing here was published as ch. 15 of *Decision, Probability and Utility,* edited by P. Gärdenfors and N.-E. Sahlin, Cambridge: Cambridge University Press, 287-312.
- "Consensus as Shared Agreement and Outcome of Inquiry" (Chapter 7) is reprinted from *Synthese* 60 (1985), 3-11, with permission of Kluwer Academic Publishers, Dordrecht, The Netherlands.
- "Compromising Bayesianism: A Plea for Indeterminacy" (Chapter 8) is reprinted from *The Journal of Statistical Planning and Inference,* 25 (1990), 347-62, with kind permission from Elsevier Science-NL, Sara Burgerhartstraat 25, 1005 KV Amsterdam, The Netherlands.
- "Pareto Unanimity and Consensus" (Chapter 9) is reprinted with permission from *The Journal of Philosophy* 87 (1990), 481-492.
- "The Paradoxes of Allais and Ellsberg" (Chapter 10) is reprinted from *Economics and Philosophy* 2 (1986), 23-53, with permission of Cambridge University Press.
- "Conflict and Inquiry" (Chapter 11) is reprinted from *Ethics* 102, 814-34, with permission of the University of Chicago Press.
- "The Ethics of Controversy" (Chapter 12) is previously unpublished.

1
Rationality and commitment

This paper discusses the function of principles of rationality in inquiry and deliberation rather than the content of such principles. Appealing to the belief-doubt model of inquiry pioneered by C. S. Peirce and J. Dewey, I shall argue that principles of rationality should impose weak constraints on the coherence of the beliefs, values and choices of deliberating and inquiring agents. Efforts to derive substantial moral or theoretical deliverances from such principles are, thereby, ruled out of court.

Weak though these constraints may be, the capacity of human and institutional agents to satisfy them is severely limited. Principles of rationality are ill suited for the prediction and explanation of human behavior. Nor can they be regarded as prescriptions which rational agents are obliged to obey to the letter. The reason is the same in both cases. Persons, institutions and other alleged specimens of rational agency lack the emotional or institutional stability, the memory and computational capacity and the freedom from self deceit required to satisfy the demands of even weak principles of coherence in belief, value and choice. Our rationality is severely 'bounded'.

In some respects, beliefs, value judgements and other so-called propositional attitudes relevant to deliberation and inquiry resemble religious vows. Just as religious vows often incur obligations only an angel could fulfil, so too, only a rational angel can satisfy the requirements imposed on rational belief, value and choice. Still, so I shall argue, the laws of heaven do have a relevance to rationality on earth. I shall try to sketch an account of what that relevance might be.

I would bet my bottom dollar that Alexander Graham Bell spent some time in Brantford, Ontario. I grant that there is a logical possibility that I am wrong; but I have no living doubt justifying my engagement in inquiry. There is no serious possibility that I am in error in any sense

Thanks are due to Akeel Bilgrami for his helpful comments and his unsuccessful efforts to save me from error.

From C. C. Gould and R. S. Cohen, eds., *Artifacts, Representations and Social Practice* (Dordrecht: Kluwer, 1994), pp. 257–75.

guiding my current conduct. To be sure, very few of the members of this audience will share my conviction. Most of them will not have heard of Brantford, Ontario, and those who, perhaps, have may be in doubt as to whether Bell ever lived there or whether he sent a telephone message from Brantford seven miles to Paris, Ontario. I myself confess uncertainty as to the precise date when this event took place and have some residual doubt as to whether this was the first long distance telephone call as is advertised by the Brantford community. Nonetheless, both those who share my views and those who differ do so by drawing a distinction between what is to be taken for granted as certainly true (or certainly false) and what is doubtful and uncertain.

Dewey thought that Peirce's distinction between what is taken for granted and what is open to question can be extended to attitudes concerning what ought to be done. Almost everyone would agree that murder is wrong. This pleasant unanimity is shattered when we ask whether abortion is wrong. At least some of the participants in the debate remain in honest doubt on the point rather than being firmly committed to either an anti-abortion or pro-choice view.

The distinction between the doubtless and the doubtful can be extended to other propositional attitudes as well. For example, an inquirer can be in doubt as to the appropriate 'prior' probability judgement to use in analysing data. For the present, however, I shall focus on the distinction between what is taken for granted in the sense of fully believed (disbelieved) and what is conjectural. I fully believe that Bell lived for a while in Brantford. That is to say, I am maximally certain that he did so. There is no serious possibility (no real and living doubt) as far as I am concerned that he did not. I remain in doubt, however, as to whether Reagan and Bush were in collusion with the Iranians over the timing of the release of hostages from the American Embassy in Teheran.

Advocates of the Peircean belief-doubt model insist that the inquiring or deliberating agent stands in no need to justify what he or she fully believes. Because I am sure that Bell lived for a while in Brantford, I lack the 'real and living doubt' required to provoke an inquiry into the matter. Justification is only required for changing one's state of full belief – i.e., modifying the distinction between what is judged to be certain and what is judged to be doubtful. Inquirers are sometimes justified in removing doubt and converting conjectures into certainties. On other occasions, inquirers are justified in doubting what they initially took for granted. In the absence of a good reason to change, however, the inquirer should retain the commitments he or she has. The first principle of pragmatism is "where it doesn't itch, don't scratch".

Thus, the fact that most readers do not share my conviction about Bell's sometime residence in Brantford is not sufficient to shake my conviction that he did reside there. However, if someone whom I respected as a competent authority on the history of southern Ontario were to weigh in with a dissent, that would normally qualify as a good reason for my coming to doubt my initial conviction and induce me to inquire further as to whether Bell did or did not live for a brief period in Brantford.

To be sure, if someone in the audience who was not the sort of competent authority I envisage demanded reasons for my convictions, I might try to oblige. But in so doing, I would not be attempting to offer reasons justifying my full belief. I have no real and living doubt which would require me to have such reasons. My attempt to offer reasons would really be an effort to explain to my interrogator why he or she ought to come to full belief that Bell lived in Brantford. I would marshall reasons which would justify the interrogator from the interrogator's point of view coming to full belief that Bell lived there for a while. But justifying a claim to another is different from justifying a claim to oneself.

I do not mean to suggest that when an agent lacks a good reason to question the agent's full beliefs, the agent is justified in suppressing the views of those with whom the agent disagrees. We tolerate the public expression of dissent while expressing our own contempt for it. To be sure, we are not obliged to register contempt either. Teachers are often obliged professionally to pay attention to points of view they judge to be patently false and, indeed, to pretend that their minds are open to such views even when they are not. This insincerity may be justified if it is effectively employed to induce students to critically examine their own views. And even where professional obligation does not support such dissembling, the demands of civility in discourse may. We should, however, distinguish between contemptuous toleration (both when the contempt is overt and when it is disguised for pedagogical purposes or considerations of conversational etiquette) and genuine respect for dissenting views. Respect for a dissenting view is displayed when the agent confronted with dissent recognizes a good reason for genuinely opening the agent's mind by ceasing to be convinced of the view the agent initially endorsed in opposition to the dissenter. Liberals of the sort I admire tolerate dissent but their toleration is often contemptuous. They need to have good reasons for opening their minds up so as to entertain seriously the views of dissenters. The mere presence of disagreement is not such a good reason. If it were, it would equate toleration and respect for the views of dissenters. Since there will be a dissenter for virtually every substantive view, advocates of toleration who conflate it with respect for the views of dissenters must be urging upon us the skepticism of the empty mind.

According to the vision present in the belief-doubt model, the central problem for epistemology is to give an account of intelligently conducted (i.e., rationally conducted) inquiries leading to modifications of states of full belief relative to which a distinction is made between what is settled and what is doubtful.

According to the Peirce-Dewey view of such inquiry, justifying changes of states of full belief (and, hence, of doubt) is attempting to adjust means to ends.[1] The ends of inquiries focused on obtaining information to settle some question of fact or theory may differ from and be irreducible to the goals of political, moral, economic or prudential decision making. Likewise the options faced will differ. But cognitive decision making remains a species of decision making. And like other types of decision making, it should be the outcome of intelligent or rational means-ends deliberation.

If the principles of rationality regulating scientific inquiry as well as moral, political, economic, religious, aesthetic and self interested deliberation are the same, the principles of rational choice needed would have

1 Consider, for example, Dewey's notion of warranted assertibility (Dewey, 1938, p. 120; 1977, p. 271).

The inquiring agent seeks to answer an as yet unsettled question. When an answer is obtained at the conclusion of such an inquiry, it is a "judgement" in Dewey's terminology. The conjectures that constitute potential solutions of the problem as well as the settled background information, techniques and methods taken as noncontroversial resources in the context of the specific inquiry are called "propositions". These are the entertainable means for realizing the given end of solving the problem. For Dewey, propositions are "affirmed" and judgements "asserted". When the inquiry is properly conducted with the appropriate adaptation of means to ends, Dewey called the assertion or judgement which represents the solution to the problem "warranted". The judgement that is the warranted assertion of that inquiry may then become a resource for subsequent deliberation and inquiry. *Qua* means of the new inquiry, it is not a warranted assertion. It is an affirmed proposition. This does not mean that it becomes a conjecture. Unless there is good reason to do so, once the inquirer has become certain at the end of the first inquiry, he or she should not cease being certain subsequently. What it does imply is that the settled assumption is now a means for realizing new ends in new inquiries.

My point in elaborating on Dewey's categories of assertion of judgements and affirmation of propositions is not to endorse their usefulness. I myself have not found them convenient. But they reveal Dewey's conviction that even scientific inquiry is a goal directed activity exhibiting features in common with technological, economic, moral, prudential and political deliberation. In this sense, justifiable change in state of belief is a species of rational decision making.

to be extremely weak. We could not require such principles to be substantial enough to support a morality or a political super-structure or, for that matter, to sustain some strong network of principles for determining how well scientific conjectures are confirmed on the basis of given evidence.

Nonetheless, the principles, no matter how weak, not only will be designed to constrain the deliberating agent's choices among options but will also presuppose for their applicability that the deliberating agent has certain kinds of attitudes. Thus, it is widely taken for granted that principles of rational choice recommending the maximization of expected utility presuppose that the agent's beliefs can be factored into the agent's full beliefs and credal probability judgements. Those who are skeptical of the propriety of expected utility maximization often question whether agents have credal probability judgements. They may, however, substitute other attitudes as expressions of uncertainty.

I shall follow familiar views according to which proposals for principles of rational choice presuppose that the deliberating agent to whom they are applied is not only in a state of full belief relative to which he or she distinguishes between settled assumptions and conjectures, but also in a state of credal probability judgement and in a state of value commitment (relative to his or her state of full belief). The state of credal probability judgement furnishes a fine grained discrimination between the conjectures with respect to the agent's judgements of credal probability. The state of value commitment assesses the conjectures with respect to the agent's values, goals and desires. The state of full belief, credal probability judgement and value commitment together with the principles of rationality determine which of the options available to the agent are admissible for him or her to choose.

The trichotomy between states of full belief, states of credal probability judgement and states of value commitment is doubtless far too coarse grained to do justice to our attitudinal life insofar as it impinges on our deliberations. However, even this rough and ready distinction can be used to emphasize another important feature of the approach favored here. An account of rational deliberation needs to impose some constraints not only on criteria for what are admissible options but also on states of full belief, credal states and value commitments. Thus, if our theory of rational choice requires that full beliefs be closed under logical consequence, that our credal state obey requirements of the calculus of probabilities and that our preferences should be transitive and obey a so-called independence axiom, these requirements are to be enshrined as principles of rational full belief, credal probability judgement and value judgement.

The principles of rationality thereby serve a dual function according to the belief-doubt model as I am reconstructing it:

1. They specify necessary conditions for a rational agent's being in a state of full belief (probability judgement, value commitment) at a given time. But since they are weak constraints, agents can change from one such state satisfying the requirements of rationality to another. Hence, the principles of rationality also circumscribe the space of potential changes in states of full belief (probability judgement, value commitment).
2. According to the belief-doubt model, changes in states of full belief (credal probability judgement, value commitment) call for justification. Since changes in states of full belief (credal probability judgement, value commitment) are to be justified by showing that such changes promote the ends of the inquiring or deliberating agent, the principles of rational choice specify minimal conditions such justification should meet.

The principles of rational choice, full belief, credal probability judgement and value judgement are neither explanatory nor predictive. No one is able to recognize as true all the logical consequences of what he or she fully believes, to fully obey the calculus of probabilities or to secure preferences (value comparisons) that are transitive. Nor can anyone even manage to approximate these requirements to any satisfactory degree except in relatively uncomplicated deliberations. Poor memory, limited computational capacity within the time constraints imposed for solving a problem, excessive costs, emotional stress, mental disease or lack of self knowledge can all conspire to prevent fulfilment of even very thin principles of rationality. So we cannot hope to use a belief-desire model of explanation to explain or predict human conduct relying on principles of rationality as covering laws. Nor is it to be expected that belief-desire models relying on qualified or modified versions of the principles of rationality will do any better. The likely outcome of inquiry aimed at identifying the requisite qualifications will be the replacement of presuppositions about the kinds of attitudes in the unmodified principles of rationality with presuppositions concerning other attitudes. The new theory of rationality is not going to do better in prediction and explanation from covering laws than the old. Those who retain a faith in principles of rationality as explanatory and predictive principles may continue to pursue their efforts to vindicate their faith in the face of the daunting obstacles they face. As far as I am concerned, the handwriting is on the wall.[2]

2 My skepticism about the use of principles of rationality as covering laws in explanation of physical behavior does not derive from doubts about the availability of psychophysical laws covering the causal interactions between mental and physical events as does Donald Davidson's skepticism about psychophysical explanation and prediction (1980, essays 11 and 12). I contend that the principles of rational belief, desire and choice fail to contribute to explanation of behavior as physically described because such principles are false as applied to human beings. It is quite clear that human and institutional agents lack and will always lack the computational capacity and emotional or institutional stability to conform to principles of rationality when the predicaments faced are too stressful or complex. I can at best assent to the truth of finitely many sentential representations of consequences of my beliefs. There are infinitely many others I could not assent to even if asked.

Thus, principles of rationality could not be "constitutive" of what is to be a rational agent as Davidson (1980, p. 221) appears to say because they are known to be false of agents. Of course, Davidson acknowledges that they are false of agents. But he does say that we cannot coherently regard someone who is subject to "global confusion" as an agent. I am not sure what "freedom from global confusion" means for Davidson. But I suspect that Davidson means that agents free of global confusion behave rationally most of the time. But whether the agent is free of global confusion in this sense would depend upon how frequently he or she faced highly complicated problems which the agent lacked the computational capacity to handle. Someone whom we would regard as a clear headed and highly intelligent agent might be thrust into a position of responsibility where his or her daily fare became complex problem solving transcending the agent's capacity. Would the erstwhile agent then cease to be an agent? Perhaps, Davidson will be able to find a satisfactory explication of the confusing notion of freedom from global confusion. But it is clear that universal conformity with the principles of rationality is not constitutive of agency. Nor are these principles covering laws over the domain of agents. Hence, they cannot be used in explanations of choices or other intentional behavior or of physical behavior. This is so whether or not there are psychophysical laws.

It may, perhaps, be objected that the principles of rationality are covering laws when a *ceteris paribus* clause is added. Those who invoke *ceteris paribus* clauses must be thinking of laws as represented by two components: (1) a domain or scope specification and (b) a formula or principle. It is not required that the distinction between principle and domain of applicability should be context independent. A lawlike universal generalization of the form "All A's are B's" can be parsed into the components (a) and (b) in at least two ways. According to one method, the scope specification is the range of the variable bound by the universal quantifier in "$(x)(Ax \rightarrow Bx)$" and the formula is the open sentence or predicate "$Ax \rightarrow Bx$". According to another, the domain is the set of A's and the formula is "Bx". In making repairs to a putative law (one that is subject to refutation by counterinstance by imposing restrictions on the domain), one often has to make a decision on the basis of what aspect of the putative law one seeks to save. If one finds an A which is not a B and seeks to identify more accurately the domain in which B holds, "Bx" becomes the formula. But if one seeks a more accurate specification of the domain in which "$Ax \rightarrow Bx$" holds, the formula is "$Ax \rightarrow Bx$". No matter how one proceeds, introducing a *ceteris paribus* clause is to offer a promissory note for restricting the domain of application of the formula or principle.

As defenders of the applicability of psychological principles in explanation and prediction of behavior are fond of pointing out, *ceteris paribus* qualifications are often found in the natural sciences as well as in psychology. However, they are promissory

Yet, I do not think that we will or should stop using belief-desire models or the principles of rationality that accompany them for other purposes. Principles of rational choice have prescriptive uses. In particular, they are often intended for use by deliberating agents for the purpose of policing their own deliberations as well as those of others.

In inquiry, such self policing is of crucial importance. The adjustment of means to ends requires the deliberating agent to identify both his or her ends and the potential means (the available options) for realizing these ends and then to determine which of the available options are admissible relative to the ends recognized. That is to say, the deliberating agent will need to identify his or her state of full belief, credal state and value commitments. Without doing so, the agent will not be in a position to apply the principles of rational choice to identify the set of admissible options.

Someone will no doubt object that the same considerations that argue against using belief-desire models for explanation and prediction pose obstacles to their prescriptive use at least by human beings and their institutions. 'Ought' presupposes 'can'. If we cannot fulfil the requirements of thin principles of rationality, it would be foolhardy and irrelevant to impose an obligation upon us to do so.

So some will argue that we should recognize that our rationality is 'bounded' and should tailor our principles of rationality to our capacities to satisfy them. But no matter how we trim our principles of rationality, there will always be predicaments so complex and so stressful as to preclude the applicability of the eviscerated standard. Evisceration will continue until nothing of interest is left to carve out.

notes which ought to be redeemed by supplying more adequate scope restrictions. In the natural sciences, the responsibility is often undertaken and often yields the result that no adequate replacement of the *ceteris paribus* clause can be found without modifying the formula. Moreover, even the restrictions on the scope and the modification of the principle initially designed to characterize complex systems typically end up being stated within the framework of some more fundamental theory. There is no *a priori* necessity that this be so; but it is quite likely to happen if *ceteris paribus* clauses attached to principles of rationality construed as components of covering laws are redeemed. The restrictions of scope will quite likely be most accurately described in biological or biophysical terms and the principles of rationality will cease being constituents of the formula.

I suppose that one might say that the principles of rationality can be ingredients in covering laws where the domain is the domain of rational, logically omniscient angels. But that set is the empty set and the covering laws for that domain are not of great interest unless other agents closely approximate the behavior of rational angels. Neither corporate nor human agents begin to come close to approximating rational angels.

In sum, the principles of rationality are neither constitutive of nor covering laws for the domain of personal or corporate agents.

Still more importantly, the counsel of those who urge us to trim our principles of rationality is the counsel of complacency. Of course, we cannot be obliged to do at the moment what we cannot at that moment do. We cannot be obliged to recognize all the logical consequences of our full beliefs or even enough of the consequences to solve some particular complicated problem. But we can be urged (costs and time permitting) to seek therapy for our distress, to devise prostheses to extend our computational capacities and memories (such as computers, paper and pencil, handbooks of tables, etc.) and to learn logic and mathematics so that we can to some extent overcome our disabilities. Authors who counsel us, instead, to modify the principles of rationality so that we can all be OK, neglect the importance of trying to enhance capacity to fulfill the demands of rationality.

Thus, we need a way of understanding the prescriptive force of the principles of rationality which can recognize our failings as rational agents while avoiding the temptation to convert such recognition into a complacent tolerance of these failings.

The worries just raised concerning the applicability of prescriptions of rational choice are also worries about whether such agents can be in states of full belief, credal probability judgement and value judgement meeting standards of synchronic rationality. This worry is different from the concern about the rational or intelligent conduct of inquiries aimed at improving states of full belief (probability judgement, value judgement) where these states are already assumed to satisfy the requirements of rationality. Nonetheless, the concern about applicability does threaten the belief-doubt model as I am construing it. The belief-doubt model offers an account of conditions under which changes in states of full belief (probability judgement, value judgement) are justified that presupposes that such states satisfy the requirements of synchronic rationality. Consequently, the belief-doubt model may be deemed irrelevant to deliberation and inquiry because no agent is ever in such a state.

We need to determine whether there is an important and useful sense in which flesh and blood human agents and the corporate agents with which they have relations can be said to be in states of full belief (probability judgement and value judgement) satisfying the demands of rationality even though we acknowledge our limited capacities for computation and recovery of information or lack of self knowledge and our psychological infirmities. And we need to show that the attitudinal states so identified are appropriate objects for critical review according to the belief-doubt model.

For the sake of simplicity, let us focus on changes in states of full belief.

Much of what I have to say can be transferred *mutatis mutandis* to credal states and states of value commitment:

Consider the following scenario:

Mario took for granted at time t_0 that Albany is north of New York City. He also took for granted at that time that 'is north of' is transitive.

Without any alteration in these beliefs, at time t_1 he also came to fully believe that Montreal is north of Albany - perhaps, as a result of looking up the location of Montreal in an atlas. There is a sense in which Mario simultaneously came to fully believe that Montreal is north of New York. If Mario were offered a bet at t_1 as to whether Montreal is north of New York where he wins $1 if Montreal is north of New York and loses nothing if it is not, he should be prepared to pay up to $1 for the opportunity to take the bet. The reason is that it is inconsistent with what he already believes that Montreal is not north of New York and, hence, there is no serious possibility that Montreal is not north of New York. The bet insures him of a sure $1 and if he thinks more money is better than less, he should be prepared to ante up any sum of money up to $1 for the opportunity.

Yet, Mario may not consciously or explicitly recognize that Montreal is north of New York and in this sense fails to fully believe that Montreal is north of New York. There are several reasons why this may not be so.

a. Mario may not have asked himself nor have been confronted by others with the question as to whether Montreal is north of New York.
b. He may have been asked the question and answered in the negative due to emotional stress, limited computational capacity or the time required to reach an answer. Or he may have been offered the bet mentioned above for an arbitrarily small positive charge and refused it.
c. He may have been asked the question and pleaded ignorance. He failed to identify the consequences of his full beliefs. Or he may have been offered the bet at various positive prices and refused it for some while accepting it for others. Perhaps again emotional stress, limited computational capacity or the time required to reach an answer prevented his putting two and two together.

Variant (a) of Mario's predicament at t_1 has been commonly handled by distinguishing between belief as a disposition to assent upon interrogation and belief as assent. More generally, when it is acknowledged that belief is a disposition not only to linguistic behavior such as assent but also to various forms of nonlinguistic behavior such as accepting bets, distinctions are recognized between belief as disposition and belief as manifestation of the disposition. If Mario were offered the bet at t_2, his disposition to accept the bet for a price up to $1 is, by hypothesis, already present. The stimulus is present and so is the manifestation. That is to say, he accepts the bet at the appropriate terms. So the difference between the change from t_0 to t_1 and the change from t_1 to t_2 is a difference between acquiring a doxastic disposition and being provoked to display it.

The contrast between belief as doxastic disposition and belief as doxastic manifestation does not accommodate predicaments (b) and (c) so well.

If at t_1, Mario had the requisite doxastic dispositions and had been interrogated, he should have responded by assenting to 'Montreal is north of New York'. In scenario (b), however, he refused the bet even for a small positive charge. Mario does not have the doxastic disposition to behave characteristic of full belief that Montreal is north of New York at t_1 according to the dispositional analysis. Yet given the full beliefs he by hypothesis does have, he should rationally have had the disposition. His failure to have it is a defect in Mario viewed as a rational agent. The remedy for the defect is to be achieved by alleviating Mario's emotional distress (or whatever) and improving his capacity to put two and two together and recognize the logical implications of his full beliefs. The remedy calls for therapy - not inquiry.

Suppose that Mario overcomes the defect at t_2. He acquires the doxastic dispositions he should already have had. There is a change in his beliefs understood as changes in doxastic dispositions. But there is no change in the doxastic dispositions he should, as a rational agent, have. That remains as it was at t_1. This stands in marked contrast to the shift from t_0 to t_1 which involves both a change in Mario's doxastic dispositions and in the doxastic dispositions he should rationally have.

Case (c) is like case (b) except for one point. Even though at t_1 Mario is interrogated, Mario pleads ignorance. This means, of course, that Mario cannot be said to fully believe as doxastic disposition that Montreal is north of New York or to manifest such a disposition. On the other hand, we have no grounds for attributing to him a doxastic disposition to dissent from 'Montreal is north of New York' either. And when offered the bet at varying prices, he displays that he is in a state of uncertainty or doubt as revealed by his accepting the bet at some prices but not at others. As in case (b), Mario has failed to have the doxastic dispositions which a rational agent in his circumstances should have. At t_2, he overcomes the failing.

I do not wish to quarrel with the naturalist view that there is a contrast between belief as a kind of 'multitrack' disposition and belief as a manifestation of such a disposition. But the difference between Mario's failure to believe that New York is north of Montreal at time t_0 and at time t_1 is not a difference in disposition. And it is not the difference between failing to display an already present disposition and manifesting it. At both times he lacked belief in the sense of a doxastic disposition to assent or to behave 'as if' New York is north of Montreal. At both times, he failed to manifest the doxastic disposition he did not have.

Why, however, is this difference between Mario's beliefs at the two times important? Mario's failure at t_1 to believe in the dispositional sense that Montreal is north of New York is a failure to do what he ought to have done. He had no such obligation at t_0.

The question still remains: How can we criticise Mario for failing to live up to a standard he cannot satisfy?

Imagine that Mario had a logically omniscient, perfectly rational *doppelgänger*, Mario's rational angel. Angelic Mario, though logically omniscient, is not omniscient. Hence, at t_0, angelic Mario has all the full beliefs flesh and blood Mario has plus all their logical consequences. At t_1, both angelic and flesh-and-blood Mario add 'Montreal is north of Albany'. Angelic Mario adds all logical consequences including 'Montreal is north of New York' as well. There is no change between angelic Mario's state at t_1 and t_2 except, perhaps, in terms of manifestations of his state as in the type (a) scenario for flesh-and-blood Mario.

We might say that Mario has an obligation as a rational though flesh-and-blood agent to mimic angelic Mario in his beliefs. Insofar as he cannot and costs permit, he should undergo therapy and training for improving his performance. The prescriptions concerning rational full belief (that it be closed under logical consequence and consistent) can be understood as regulating the way obligations are generated while providing for excuses due to human computational and emotional incapacity.

This does not as yet determine whether angelic Mario was justified in shifting from the doxastic dispositions he had at t_0 to those he had at t_1 or whether the flesh-and-blood Mario was justified in altering his obligations to acquire doxastic obligations accordingly. This issue is not a question for therapy or technology to handle. The justification is obtained at the outcome of inquiry. Unlike angelic Mario, flesh-and-blood Mario stands in need of therapy at t_1. The change in his beliefs from time t_1 to time t_2 is a product of such therapy.

Abandoning the myth of the angelic Mario, from whence comes this doxastic obligation which in scenarios (b) and (c) Mario not only fails to fulfil but is not able to fulfil at t_1?

Clearly, whatever we want to say about the source of Mario's doxastic obligation, we would have to locate it in his adding full belief that Montreal is north of Albany in the dispositional sense to what he already fully believed at t_0. That act of adding new information to the initial belief state generated new doxastic obligations stronger than those he already had.

The idea that doxastic obligations are generated by doxastic dispositions suggests that changes in full belief illustrated by Mario's change from t_0 to t_1 are really changes in commitment (and not merely in disposition) in a sense I shall now try to explain.

Undertakings or commitments resemble contracts and promises in three respects: (1) Making a promise or drawing up a contract requires the occurrence of events describable in physical terms. (2) These events are redescribable as promises or contracts only if their occurrence is understood as generating obligations. (3) Prescriptive principles and their interpretations need to be invoked in order to construe the event described in physical terms as generating obligations. The putting of pen to paper must be interpreted as signing a contract to pay a sum of money by a certain time and such signing of contracts must be seen as generating an obligation to pay the sum of money by the due date. The obligation is generated by the signature in virtue of the laws of contract or moral principles invoked.

Mario's change in state of full belief from t_0 to t_1 resembles signing a contract in these respects precisely. Mario's change in state of full belief is a change not merely in some doxastic dispositions but also and more importantly in doxastic commitment. In acquiring the doxastic disposition to assent to 'Montreal is north of Albany', Mario has undertaken a commitment to maintain that doxastic disposition in addition to the doxastic dispositions he already had at t_0 pending the acquisition of good reasons for altering them. His undertaking generates further commitments to acquiring doxastic dispositions to assent to the logical consequences of 'Montreal is north of Albany' and the doxastic dispositions he had at t_0. In particular, Mario undertook to be disposed to assent to the truth of the claim that Montreal is north of New York. In lieu of the law of contract, the principles of rational belief generate the obligations.

Thus, we have a three way polyguity in our understanding of 'full belief': (A) full belief as doxastic commitment; (B) full belief as doxastic disposition – I shall call this full recognition rather than full belief; and (C) full belief as manifestation of doxastic disposition – I shall call this manifestation of recognition or doxastic disposition.

Full belief as doxastic disposition and as doxastic manifestation has unavoidably prescriptive and evaluative components. The behavioral or physical dispositions involved are interpreted as fulfilling a doxastic obligation or as generating such obligations. This interpretation cannot be obtained by appeal to information about the natural world alone.

But this ought not be so surprising. That is what Brentano's thesis has already told us about the gap between the intensional and the natural. The only novelty introduced, if it is a novelty, is that the irreducibility of the intensional to the physical is itself equated with the irreducibility of the prescriptive to the physical. More would-be naturalists may be prepared to resign themselves to the intractability of the fact-value dichotomy than to the gulf between the intensional and the natural.

We have not, however, fully addressed the difficulties about lack of logical omniscience which have provoked many commentators to follow H. A. Simon in trimming the standards of rationality. When one makes a promise one is certain one cannot keep, one's conduct is fraudulent. By comparing belief states to undertakings where the undertakings generate obligations extending well beyond our capacities, belief states are construed as fraudulent commitments.

Doxastic commitments are not contracts or promises. But there are other kinds of undertakings. Religious vows are often seen as generating undertakings well beyond the ability of the person taking the vows to satisfy. Someone vows to be righteous knowing full well that he will never satisfy all the requirements of righteousness. Yet such vows are not fraudulent.

The reason they are not fraudulent is that they do not, strictly speaking, obligate the one who undertakes the commitment to meet them fully or even approximately but only insofar as the agent has the capacity to do so. However, this rather substantial qualification of the obligation undertaken is accompanied by another substantial obligation. The agent is required to extend his or her ability to fulfil the obligations by training and education, the use of appropriate therapies, and the acquisition of prosthetic devices enhancing his or her capacities. There is no fraud because the person who has undertaken the religious vows is not obliged to do what he or she cannot do although he or she is obliged to seek ways and means to enhance his or her capacities.

Doxastic commitments resemble such religious undertakings in this respect. We are committed doxastically to fully believe the consequences of what we fully believe and to believe that we believe them. This does not mean that Mario at t_1 is obligated to fully recognize that Montreal is north of New York even when, as in scenarios (b) and (c) he cannot do so. However, given that he cannot do so, he should seek ways and means to enhance his capacity to put two and two together.

Thus, the inquiring and deliberating agent is committed to improving the extent to which he or she meets the demands of logical omniscience by studying logic and mathematics, using computers and other prosthetic devices and undergoing forms of psychotherapy.

Suppose then at time t_2, Mario, who at t_1 is in either version (b) or (c) of the predicament, does put 2 and 2 together and successfully recognizes that Montreal is north of New York. The change from t_1 to t_2 is clearly different from the change from t_0 to t_1. The latter change is a change in Mario's doxastic commitment and is best seen as the outcome of inquiry. The former change involves no change in commitment but only in what I shall call 'doxastic performance'. And it is a result of therapy or the use of prostheses.

Version (a) of Mario's predicament at t_1 is slightly different. Here Mario is not tested at t_1. It is left unsettled as to whether Mario has the doxastic dispositions fulfilling his doxastic commitments or does not. If the former is the case, the shift at t_2 is one where there is no change either in doxastic commitment or in doxastic performance but only in the manifestation of Mario's doxastic performance. In the latter case, there is a change in performance.

So far all of my examples and discussion have been about full belief. However, the contrasts I am sketching here apply *mutatis mutandis* to probability judgement and value judgement. On the view I am proposing, principles of rational full belief, credal probability judgement and value judgement characterize the way in which states of full belief, credal probability judgement and value judgement understood as states of attitudinal commitment generate obligations to full recognitions, conscious probability and conscious value judgements that only rational angels could fulfil perfectly. We are not obligated so strictly. God is merciful. We are, however, obliged to improve our capacities to fulfil our commitments, opportunities and costs permitting.

I first bruited an analogy between belief as commitments and promises or contracts. I admitted the analogy was only partial and introduced a second analogy between belief as commitment and religious vows which overcomes some of the defects in the analogy with promises. But this analogy too may seem only partial. Someone might ask: To whom has Mario made his doxastic vows when undertaking doxastic commitments?

And what are the sanctions legitimizing the prescriptions of rational full belief, credal probability and value judgement which generate obligations from performances?

Dewey insisted that one can behave religiously without being committed to any sort of theology (Dewey, 1934). Atheists, agnostics and other secularists can behave religiously. I would go further. We need not suppose that if the vows are not to God they must be to some community or individual. The vows are to no one or no one group. And the ends we take for granted do not require justification as long as they are taken for granted.

I am a secularist. It is enough for me that principles of rationality can be understood as characterizing the commitments undertaken in states of full belief, probability judgement and value judgement. With this understood, we can distinguish between change in belief (probability judgement, value judgement) construed as change in performance and best effectuated by therapy and training and change in belief (probability judgement, value judgement) construed as change in commitment.

I suggest that the province of inquiry according to the belief-doubt model of Peirce and Dewey is to rationalize changes in attitudinal commitments whereas therapy, technology and education address the improvement of performance. Clearly one cannot have the one kind of change without the other. Doxastic performances cannot be recognized to be performances unless they are understood as attempts to realize commitments. Understanding comes when the commitments which the agent is attempting to fulfil are recognized. Commitment without performance is empty. But we should also remember that performance without commitment is blind.

Principles of rationality, therefore, have two tasks to perform: They regulate changes in commitment by furnishing the criteria for admissibility that warrant choosing one change in commitment over another. And they regulate changes in performance by indicating the standards to which performance would conform were we rational angels and thereby furnish the ideals of rational health which should guide our psychotherapies, our educational programs and our technologies.

Both Dewey's vision of the intelligent adjustment of means to ends and my view of principles of rationality as weak constraints on rational full belief, probability judgement, value judgement and choice are paradigmatically conceptions of instrumental rationality. We both contend that it is rationality so construed that is central to justifying changes in doxastic and other attitudinal commitments (such as changes in probability and value judgement as the first task requires).

Nonetheless, these principles also constrain our interpretation of the propositional attitudes as dispositions – i.e., as performances and as manifestations fulfilling or failing commitments as the second task demands. In performing this task of characterizing the standards of rational health, the values embodied are themselves not to be rationalized by reference to other goals and values they promote. Although they are secular in the sense they entail no theology, they have the features of religious commitments I have sketched above.

This brings me to the important respect in which the view of rationality I have been outlining may differ from Dewey's. Dewey seems to have held that our full beliefs in the truths of logic are liable to reasoned change. I surmise he would have said the same about commitments to minimal requirements on the coherence of probability judgement and value judgement as well. Nothing is fixed in concrete including the thin principles of rationality.

In one respect, I share Dewey's view. Weak principles of rationality characterizing our doxastic and other attitudinal commitments have changed historically. Even at the present, they are subject to controversy and change.[3] The problem is to give an account of good reasons for making such alterations. Such an account would have to meet two requirements: (a) It would have to be comprehensive enough to address good reasons for changing full beliefs, probability judgements and value judgements and (b) it would need to avoid begging controversial questions about changes in the standards of what constitutes good reasons. I do not know how to meet these demands.

Nonetheless, there are certain considerations that are relevant in assessing standards of rationality. I have already invoked them in this discussion. The standards should be strong enough to perform their functions in interpreting the attitudes as commitments and performances and evaluating choices in the light of such interpretations. Yet, they should be weak enough to accommodate a broad spectrum of belief states and value commitments so as to recognize such states and commitments as potentially subject to reasoned review and alteration as is feasible.

In opposition to Kant and other enlightenment figures, we should try to avoid rendering our favorite substantive doxastic and evaluative commitments secure against reasoned change by seeing them as dictates of rationality. Morality and science are one thing. Rationality is quite another. We may expect our morality and our science to conform to the

3 I myself championed substantial changes in ideals of rational choice envisaged by strict Bayesians and other students of rational choice (Levi, 1974, 1980, 1986).

requirements of rationality. But we should resist the pretensions of those who seek to ossify morals or science in the name of reason or, who, while acknowledging the changeability of morals and science, see such variability as manifesting a variability in the standards of rationality that cannot be rationalized.

To the extent that we can meet the two vaguely specified demands mentioned above, we may hope to come close to avoiding these pitfalls and fulfilling the spirit if not the letter of Dewey's view.

REFERENCES

Davidson, D. (1980), *Essays on Actions and Events,* Oxford: Clarendon Press.

Dewey, J. (1934), *A Common Faith,* New Haven: Yale University Press.

(1938), *Logic: The Theory of Inquiry,* New York: Holt.

(1977), *Dewey and His Critics,* ed. by S. Morgenbesser, New York: *Journal of Philosophy.*

Levi, I. (1974), 'On Indeterminate Probabilities,' *Journal of Philosophy 71,* 391-418.

(1980), *The Enterprise of Knowledge,* Cambridge, Mass.: MIT.

(1986), *Hard Choices,* Cambridge: Cambridge U. Press.

2

Rationality, prediction, and autonomous choice

Principles of rationality are invoked for several purposes: they are often deployed in explanation and prediction; they are also used to set standards for rational health for deliberating agents or to furnish blueprints for rational automata; and they are intended as guides to perplexed decision makers seeking to regulate their own attitudes and conduct. These purposes are quite different. It is far from obvious that what serves well in one capacity will do so in another. Indeed, I shall argue later on in this essay that when principles of rationality are intended for use as norms for self-criticism, they cannot also serve as laws in explanation and prediction or as blueprints for rational automata.

Whether such principles are used for explanatory purposes, for setting standards or for self policing, issues arise as to the scope of applicability of principles of rationality. Such principles are often employed to evaluate decisions as well as the attitudes which allegedly inform them. Whether principles of rationality are relatively weak constraints of 'coherence' and 'consistency' on belief, desire, probability judgment and other propositional attitudes is by no means settled. Many writers argue to the contrary. They insist that norms of rationality should be 'thickened' to include more substantive specifications of the beliefs, values, probability judgments which agents should have. At the thicker end of the spectrum, for example, is found the idea that morality and politics should somehow be derived from a conception of rationality, that factual beliefs and probability judgments should be rationally determined by 'evidence.'

The question of scope raises issues not only about how strong principles of rationality should be but also about the domain of applicability. Many sympathizers with the Humean perspective doubt that 'the passions' can or should be subjected to even thin constraints of rationality in the manner in which it is alleged that beliefs can be criticized with respect to consistency.

Thanks are due to Carole Rovane for shrewd and patient criticism on matters of substance and style.

From *Canadian Journal of Philosophy,* Supp. Vol. 19 (1989), pp. 339–63. Reprinted by permission of The University of Calgary Press.

In the face of the many possible positions which can be taken on these matters, proposing an account of *the* concept of rationality is either foolhardy or presumptuous. I prefer instead to explain my motives for focusing on one kind of understanding of rationality.

My interest in rationality (or whatever one may wish to call it) derives from a fascination with the Peirce-Dewey 'belief-doubt' model of inquiry. I shall first seek to explain a way of understanding principles of rationality suited to a systematic account of the core features of well conducted inquiry according to the belief-doubt model. This discussion will then serve as background for my main contention that principles of rationality designed for use in self criticism cannot simultaneously be used for explanatory and predictive purposes.

The belief-doubt model begins with a distinction between what is taken for granted (i.e., is certain or fully believed) and what is conjectural (i.e., possibly false or open to doubt). It contends that the inquiring agent is in no need to justify what he or she fully believes. Justification is required for *changing* one's state of full belief – i.e., modifying the distinction between certainty and conjecture. On the one hand, inquirers seek to remove doubt and thereby convert conjectures into certainties and, on the other hand, there are occasions where inquirers should come to doubt what they initially took for granted. In the absence of a good reason to change, the inquirer should retain the commitments he has.[1]

According to this vision, the central problem is to give an account of well conducted (i.e., rationally conducted) inquiries leading to modifications of the divide between what is settled and what is doubtful.

A central feature of the Peirce-Dewey view of such inquiry is the

1 To avoid any misunderstanding, it should be emphasized that this claim does not imply that the inquiring agent should regard himself as justified in suppressing the views of those with whom he disagrees. We may tolerate the public expression of dissent while expressing our own contempt for it. But we are not obliged to register contempt either. Indeed, the agent might sometimes listen to the dissenter's view even though those views seem absurd. Teachers are often obliged professionally to pay attention to points of view which they judge to be patently false and, indeed, to pretend that their minds are open when they are not. Such insincerity may be justified if it is effectively employed to induce students to question their views. And where professional obligation does not support such dissembling, the demands of civility in discourse may. We should, however, distinguish between contemptuous toleration both when the contempt is overt and when it is disguised for pedagogical purposes or due to considerations of conversational etiquette and genuine respect for dissenting views. Respect for a dissenting view arises when an agent confronted with dissent recognizes a good reason for genuinely opening his or her mind by ceasing to be convinced of the view initially endorsed which is in conflict with the dissent. Liberals of the sort I admire tolerate dissent but their toleration is often contemptuous. To confuse toleration with respect for the views of dissenters can lead advocates of toleration to urge upon us the skepticism of the empty mind.

assumption that justifying changes of states of full belief (and, hence, of doubt) is a species of attempting to adjust means to ends. The inquiring agent seeks to answer an as yet unsettled question. The conclusion of such an inquiry when an answer is obtained is a 'judgment' in Dewey's idiosyncratic terminology. The conjectures which constitute potential solutions of the problem under scrutiny are 'propositions' and, hence, are to be regarded as entertainable *means* to the given *end* of solving the problem. These means include not only conjectures or potential solutions to the problem but also the settled background information, techniques and methods taken as noncontroversial resources in the context of the inquiry. According to Dewey, propositions are affirmed and judgments asserted. When the inquiry is properly conducted with the appropriate adaptation of means to ends, Dewey called the assertion or judgment which represents the solution to the problem 'warranted.' The judgment which is a warranted assertion of one inquiry may then become a resource for subsequent deliberation and inquiry. Qua means of the new inquiry, it is not a warranted assertion. It is an affirmed proposition. This does not mean that it becomes a conjecture (unless called into question in the course of inquiry) but merely that background information, like conjectures, constitute ingredients of the instrumentalities of ongoing inquiry.[2]

2 John Dewey (1938, pp. 118–20, 1941, pp. 270–2).

According to Dewey, every case of knowledge 'is constituted as the outcome of some special inquiry. Hence, knowledge as an abstract term is a name for the product of competent inquiries' (Dewey, 1938, p. 8). In Levi (1983, pp. 27–8), I suggested that Dewey should have abandoned this conception of knowledge. He should not have insisted on equating knowledge with warranted assertibility if he also wished to acknowledge, as Peirce had, that to be settled, a conviction did not require a prior justification. It may, perhaps, be thought that an appreciation of the context sensitivity of Dewey's conception of 'warranted assertibility' would mitigate the difficulty. But if a proposition used as background information in an inquiry is not a warranted assertion relative to some prior inquiry, it is not knowledge even in the current inquiry in Dewey's official sense. Since Dewey is quite clear that his sense of 'knowledge' is honorific, withholding the epithet is tacit attribution of some deficiency. But it is precisely the allegation of the existence of such a deficiency that Peirce sought to rebut in 'The Fixation of Belief.' Knowledge and warranted assertibility cannot coincide as Dewey suggests. There can be propositions used as background information and, hence, qualifying as knowledge which are not products of prior properly conducted inquiries and, hence, are not warranted assertions relative to any historically conducted inquiries. Moreover, it can even happen that knowledge fails even for items that are warrantedly assertible relative to one inquiry provided the result claimed has been legitimately called into question in subsequent inquiry.

It may, perhaps, be worth mentioning that Dewey tended to think of conjectures or hypotheses as what I have tended to call 'potential answers' to the question under investigation. For me the distinction between settled assumptions and what is open to doubt or conjectural is presupposed by the deliberating or inquiring agent when facing a problem. It is one of the tasks of inquiry addressing a certain question, to identify

I do not suggest that we adopt Dewey's nonstandard and poorly understood terminology (as attested to by the way 'warranted assertibility' has been misconstrued by subsequent commentators). My point is that Dewey's categories of assertion of judgments and affirmation of propositions constituted an attempt on his part to emphasize his view that even scientific inquiry is a goal directed activity exhibiting features in common with technological, economic, moral, prudential, and political deliberation. Justifiable change in state of belief is itself a species of rational decision making. The ends of inquiries focused on obtaining information to settle some question may differ from the goals of political, moral, economic, or prudential decision making. Likewise the options faced will differ. But cognitive decision making remains a species of decision making. And this observation suggests that a concern with theories of rational choice should be of interest to those who wish to articulate a pragmatist version of the belief-doubt model.

Presumably such an account of rational decision making would have to observe a neutrality with respect to the kind of decision problem under consideration. It must be an account of rational decision making which is applicable both to cognitive decision problems and to other kinds of decision making which differ from one another with respect to both their goals and the means and options deployed.

To be sure, no articulate account of rational decision making can avoid making some general assumptions about decision problems in general. So-called Bayesian decision theories, which promote maximization of expected utility, presuppose that the deliberating agent takes for granted that he or she has a certain set of options available to him or her and that if the agent were to become certain that he or she will implement one of these options, one of a given list of conjectures as to what the relevant 'outcome' of the implementation will be is true. According to the strict Bayesian view the agent is committed to a network of credal or subjective probability judgments concerning the truth of these conjectures conditional on the option being implemented and also to an evaluation of the hypotheses about outcomes representable by a utility function.[3]

potential answers to the question under investigation. This is the task of what Peirce called 'abduction.' Three minimal demands should be imposed on a potential answer: (a) Its truth should be consistent with what is taken for granted, (b) it should be a relevant answer to the question raised, and (c) whether it is to be accepted or rejected should be decidable through inquiry. Condition (a) indicates that potential answers or conjectures in Dewey's sense presuppose a distinction between certainties and conjectures in my sense.

3 The utility function representation is to be understood in this setting as characterizing the agent's values and goals in the deliberation and not as representing the net of pleasure over pain of his or her desires. The utility function could represent moral,

I am not concerned for the present with the technical details of this Bayesian vision of rational decision making. Nor, for that matter, am I in favor of this Bayesian vision without substantial qualification.[4] My point in mentioning it is that it illustrates a feature of such weak conceptions of rational choice. There is not only a need for principles for evaluating the deliberator's options but also for conditions of coherence or consistency on the judgments as to which options are feasible, what the possible consequences of a given option are, the credal probability judgments and the utility judgments. The inquiring or deliberating agent is assumed to be drawing not only a distinction between what is certain and what is conjectural but finer grained discriminations among the conjectures with respect to credal probability and utility. We need to identify the constraints on rationality imposed on these judgments. To this extent, therefore, we need, in Alan Gibbard's words, an account of not only 'wise choices' but 'apt feelings.'[5]

We need to do more than this. The credal probability and utility judgments of agents, like the distinction between certainty and conjecture, are subject to change. A sophisticated version of the belief-doubt model of inquiry should provide an account of when and how judgments of credal probability and of value should be revised in the course of deliberation and inquiry. And this suggests that we should provide an account of a distinction between what is settled and what is doubtful in probability and utility judgment.

Thus, Charles Peirce was vociferously skeptical of the Bayesian idea that rational agents should have numerically definite credal probability judgments.[6] Unless such judgments are suitably grounded in knowledge of objective probabilities or chances, he thought they were doubtful. John Dewey insisted that doubt concerning what is for the best is a ubiquitous feature of our moral predicament and called for an extension of the belief-doubt model into questions of value.[7]

If one is going to give an account of rational decision making, rational full belief, rational probability judgment and rational value or utility judgment which offers any hope of allowing for a viable account of the budget of questions just thrown out for consideration, it is to be expected

political or cognitive valuations. The core principle involved is the principle of maximizing expected utility conditional on the act chosen. See Levi (1980, ch. 4) for a survey of core features of Bayesianism.

4 For further elaboration of these qualifications, see Levi (1980 and 1986).

5 Alan Gibbard (1990).

6 See C. S. Peirce (1984-86, v. 2, pp. 98-102, v. 3, pp. 300-1) and Levi (1991a, pp. 99-103).

7 See John Dewey and John H. Tufts (1932); and Levi (1986, ch. 1).

that the principles of rationality regulating these various propositional attitudes should be weak. The reason is that the account of change of view (whether it is change in what is counted as certain, what is judged probable or valuable) will depend on holding these principles of rationality fixed. Perhaps there is some deep sense in which no principles are to be held constant; but if one is seeking to give a systematic account of deliberation and inquiry, relative to that account some principles are going to be immune to revision. If the account is to be supple enough to provide a robust account of deliberation and inquiry, the fixed principles of coherent or consistent choice, belief, desire, etc. will have to be weak enough to accommodate a wide spectrum of potential changes in point of view. We may not be able to avoid some fixed principles, but they should be as weak as we can make them while still accommodating the demand for a systematic account.

To this extent, the view of rationality I am deploying departs from Gibbard's conception according to which 'to call something rational is to express one's acceptance of norms that permit it' (1990, p. 7). Granted that acceptance of norms of coherence or consistency controls the rational permissibility of 'choices' and 'feelings.' But in giving an account of the rational revision of norms through inquiry as Dewey sought to do, some sort of distinction needs to be made between the norms open to revision and those which are taken to characterize the rational conduct of inquiries concerned with such revision. Dewey thought, for example, that moral inquiry is just that kind of concern so that the moral principles which are the objects of criticism cannot be norms characterizing the rationality of the inquiry.[8]

The accounts of deliberation and inquiry offered by Peirce and Dewey were not intended to be fragments of sociology or psychology. Although Dewey insisted that certain biological and social conditions are necessary

8 It may be pointed out that Dewey himself would have regarded the distinction I am making between norms of rationality and other norms as an untenable dualism. All norms are in some context or other objects of criticism and inquiry. But if our aim is to offer a systematic account of the rational conduct of inquiries concerned with the revision of beliefs, goals, values and other attitudes, the proposed accounts we offer will perforce draw a distinction between the norms which are held fixed in these accounts and those which are open to revision. I grant that disputes can arise as to which norms should qualify as the fixed norms of rationality. But any specific proposed account of inquiry acknowledges some distinction between norms which are fixed and norms which are open to revision. Rival proposals should, if they entertain the ambition of being systematic, share in common recognition of some sort of distinction between the fixed minimal norms of coherence or consistency and norms which are open to revision even if they differ concerning how the line is to be drawn. I propose to restrict the norms of rationality relative to an account of inquiry to the fixed minimal norms specified by the theory.

for inquiry to take place and, indeed, for well conducted inquiry to take place, good methods of inquiry are distinguished from bad and the characterization of norms of method is prescriptive.

I have suggested that the weak principles of rationality to be constructed could be construed as normative standards of rational health. Alternatively, they could be deployed by deliberating agents to evaluate their options, full beliefs, probability judgments and value judgments to ascertain whether they satisfy the requirements of a weak account of coherence and consistency. That is to say, the principles of rationality should be applicable in self criticism.

Peirce and Dewey thought of the prescriptions central to the belief-doubt model as available to the inquiring or deliberating agent for the purpose of self criticism in the context of deliberation or inquiry.[9] To be sure, such norms were also to serve as standards of rational health. But individuals well educated according to such standards were presumably to be trained to think for themselves - that is to say, to be in a position to bring the standards for rational health to bear in their own deliberations. In any case, I think an account of inquiry should be characterizable by norms available for nonvacuous self criticism. Hence, the standards of rationality relevant to the discussion of the belief-doubt model should be norms of this variety.

When used for self policing, the applicability of the principles should

9 Neither Dewey nor Peirce explicitly claims this. Principles of reasoning are habits or leading principles or the like. The rhetoric, however, could often be read as blueprints for rational automata or as principles applicable in self criticism. Yet some passages suggest the latter reading fairly clearly. Thus Dewey writes: 'A postulate is also a stipulation. To engage in an inquiry is like entering into a contract. It commits the inquirer to certain conditions. A stipulation is a statement of conditions that are agreed to in the conduct of some affair. The stipulations are at first implicit in the undertaking of inquiry. As they are formally acknowledged (formulated), they become logical forms of various degrees of generality. They make definite what is involved in a demand. Every demand is a request, but not every request is a postulate. For a postulate involves the assumption of responsibilities. . . . On this account, postulates are not arbitrarily chosen. They present claims to be met in the sense in which a claim presents a title or has authority to receive due consideration' (Dewey, 1938, pp. 16-17). Dewey continues later to observe that when a specific person engages in inquiry, 'he is committed, in as far as his inquiry is genuinely such and not an insincere bluff, to stand by the results of similar inquiries by whomever conducted. 'Similar' in this phrase means inquiries that submit to the 'same conditions or postulates' (ibid., 18). The postulates Dewey is talking about here and which he regards as the terms of the contract an agent enters in undertaking an inquiry are clearly, for this reason, normative or prescriptive. As terms of a contract, they are intended to formulate prescriptions which the party to the contract endeavors to meet. When the undertakings are explicit, Dewey regards them as postulational. I take this to mean that postulates are principles the agent can explicitly recognize and use in evaluating the extent to which he is fulfilling his contract.

be nonvacuous in the sense that a nontrivial distinction may be made between feasible options which are admissible for choice and others which are not. If the principles of rational choice never eliminate any feasible option from the relevant set of feasible options, they fail to serve this function. It may still be possible to construe such principles as blueprints for designing rational automata or as principles for predicting behavior. But they will fail as standards of rational health for self critical agents and as principles for self policing.

The thesis I wish to advance is that this demand for nonvacuous self applicability entails an asymmetry between the first person perspective and the third person perspective which has no bearing on first person privileged access but which does pose a serious obstacle to viewing principles of rational choice designed to be nonvacuously applicable in self criticism as generalizations useful in prediction and explanation of human behavior. I shall argue that the asymmetry thesis does not hold if we rest content with viewing the normative principles of rationality as blueprints of conceptual, deliberative or rational health without expecting that agents use them in policing their own decisions. In that case, however, standards for rational health are blueprints for rational automata which simulate the behavior of rational agents but fail to employ the principles of rational choice to determine which feasible options are admissible. It is not clear to me that the classical pragmatists appreciated this point. As a consequence, the impression was given that emphasis on the explanatory uses of principles of rationality coheres well with the main tenets of pragmatism. In any event, many contemporary thinkers who identify themselves with pragmatism do not appear to be sensitive to the issue.

The asymmetry thesis is rich in consequences for contemporary conceptions of the mind and for theories of rational choice. In this essay, I can do no more than gesture towards what these consequences are. I shall rest content here in sharpening and defending the asymmetry.

A decision maker X engaged in deliberation needs to identify a roster of options X judges feasible for choice. Such judgments are crucial to any assessment by X of what it is rational for X to choose. X might recognize an option which, were it available to him, would be preferable to one he judges available to him. But if X judges the option unavailable to him, he is not required to judge it irrational of him to refuse to choose it rather than one of the options he recognizes as feasible. In applying criteria of rational choice to identify a set of options admissible for X to choose, X applies these criteria to a set of options which are feasible according to X.

If Sam, for example, is confronted with the choice of playing a piece by Chopin on the piano or surrendering the contents of his wallet and doubts that he has the ability to play the Chopin piece by his choice, he does not recognize playing the piece to be feasible for him. If Sam judged it feasible, he would prefer playing than paying. Failing to play would be irrational. But it is not irrational for him to pay if he cannot play.

Perhaps, the following objection will be raised: Unless Sam is certain that he is incapable of playing by choice, he has, from his point of view, the option of *trying* to play.

That may well be true. But trying to play is a different option from playing. To judge himself as having the option of playing, Sam must take for granted that his choice of playing is efficacious. He must be convinced that he will play if he chooses. If he were to have doubts about efficaciousness, he should not judge it feasible for him to play but at most to try to play. And whatever may be meant by 'trying to play' (e.g., making it objectively more probable that Sam will play), if trying to play is an option, Sam should be certain that his choosing this option will be efficacious - i.e., will render it objectively more probable that he will play. Being certain that the choice of an act is efficacious is a second necessary condition on judgments of feasibility.

Whether Sam judges that he can play the piano by choice or that he can only try, Sam is taking for granted that he has certain abilities (e.g., to play by choice). Sam's judgments of his objective abilities belong in Sam's state of full belief. Hence, Sam's state of full belief (and this holds true for any deliberating agent) must contain more than logical and other conceptual or a priori truths.

Abilities are duals of (sure fire) dispositions and, like sure fire dispositions, are relative to kinds of trials, experiments or initiating conditions. When a piece of sugar is alleged to have the sure fire disposition to dissolve in water, it is taken for granted that any water soluble thing dissolves if immersed in water. When a coin is alleged to have the ability to land heads on a toss (conditional on the coin's being tossed), it is taken for granted that anything possessing this ability lacks the sure fire disposition to fail to land heads on being tossed. Thus, if Sam has the ability to play Chopin by choice (i.e., conditional on deliberating) he lacks the sure fire disposition to fail to play Chopin by choice.

The relativity to kinds of trials or initiating conditions is of crucial importance. Sam may have the ability to play Chopin conditional on deliberating but at the very same time lack the ability to play Chopin by deliberating while suffering from an asthma attack. Sam could consistently be certain that he has the first ability and lacks the second. His conviction that he has the first ability may be necessary to his judging his choosing to play the piano to be a feasible option. But it is scarcely sufficient. If he also is certain that he is having an asthma attack while deliberating, it is not possible as far as Sam is concerned that he play the piano. That is to say, Sam is certain that he will not play the piano. Even though he has the ability at the time to play the piano conditional on deliberating and is deliberating, as far as he is concerned playing the piano is not optional for him. The extra information that he is suffering from an asthma attack precludes his coherently judging that playing is feasible for him. It is not irrational for him to surrender his wallet even though he would have wanted to play the piano rather than to pay were he facing that choice.[10]

Thus, Sam's full beliefs identify (1) what it is (objectively) possible for him to do through his deliberations (i.e., to choose to do) and (2) whether his choices are efficacious.

Given that Sam is certain that he has the objective ability to play Chopin through his choice and that his choices are efficacious, necessary conditions are satisfied for his playing Chopin to be feasible. Are these necessary conditions jointly sufficient? I think not. Feasibility presupposes that implementing the option is a 'serious possibility' – i.e., is consistent with the agent's state of full belief. If Sam is certain that he will not yield his wallet, paying is not possible as far as he is concerned. Once the matter is settled in this respect so that it is no longer consistent with what he takes for granted, feasibility is also precluded. That is to say, even if Sam would have preferred paying to playing (assuming paying was a serious possibility), it is not irrational for Sam to play given that paying is not a serious possibility relative to what he knows.

This third condition obtains regardless of how Sam came to be certain of this. Perhaps, it is because Sam has already decided to play. Perhaps, Sam's decision displayed weakness of will. He initially renounced paying while preferring to do so. His incontinence may be a form of irrationality.

10 The assumptions which the deliberating agent makes concerning his abilities resemble in certain important respects the assumptions made when an inquirer judges that a stochastic experiment is to be implemented. If a die is about to be tossed once, the inquirer presupposes that exactly one of six kinds of outcome is about to occur. These six possible outcomes or points in the sample space represent abilities of the die to respond in these six ways on a toss. The die is also presupposed to have a sure fire disposition to land in exactly one of these ways on a toss. In deliberation, the agent makes analogous assumptions. He takes for granted that he has the ability to make true each of a variety of propositions through his deliberation (through his choice) and that he is constrained by deliberation to make exactly one of these propositions true. The space of objectively feasible options is like a sample space. See Levi (1986, ch. 4).

Yet it remains just as incoherent to continue to judge the option of playing as feasible. This point has, perhaps, more bite in cases where the agent is initially faced with more than two options. Jones may have interviewed three candidates for a job and may have decided (rationally or irrationally as the case may be) to reject the third candidate. As matters stand, the only available options are to hire the first or the second candidate. Given his decision to reject the third candidate and the efficaciousness of his choices, it is not epistemically or seriously possible that he choose the third candidate as far as he is concerned. Because hiring the third candidate is not a feasible option given Sam's convictions, rationality does not require that he take that option into account in determining what to do.

It may, perhaps, be objected that Sam can renege on his past decision. *If* reneging is an option for him and *if* he is not certain that he will not renege, the point is well taken. But given that Sam has chosen to reject the third candidate under the assumption of efficaciousness, he has ruled out reneging as a serious possibility. To be sure, Sam may subsequently change his mind and conclude that his initial decision is not efficacious after all. But as long as he fails to do so, he remains certain that he will not choose the rejected option. Consequently, in the context of his deliberation at the time, the rejected option is not a feasible option for him.

Thus, whether Sam coherently judges an option as feasible for him in the context of his current deliberation depends not only on his taking for granted assumptions about his abilities and efficaciousness but on his *not* taking for granted certain other claims but rather regarding them as serious possibilities. Sam's judgment of feasibility is dependent not only on what he knows but on his ignorance as well.

Suppose then that Sam judges that both playing the Chopin and giving up his wallet are optional for him. To determine what he should do rationally, Sam needs to identify his goals and values and how they together with his full beliefs and credal probability judgments determine which of the options he judges available to him are 'admissible' - i.e., not prohibited for choice. Thus, to apply his criteria of rationality to his problem, Sam needs to access information about his goals and values, his full beliefs and credal probability judgments and his criteria of rational choice and have enough computational capacity or logical omniscience to identify which of the options is admissible or, if both are, to reach this conclusion as well.

Sam may or may not manage these feats. Confusion, emotional disturbance, self deception and the like inhibit Sam's efforts to identify his values and beliefs. And if the structure of the problem is complex enough, he may lack the computational capacity to reach a definite conclusion on the basis of the information he has succeeded in eliciting. Thus, it is clear that criteria for rational choice can fail to be self applicable and often are. This is so whether 'Bayesian' or rival criteria are used provided they carry sufficient suppleness to reflect nuances of different types of decision problem. Such sophistication is always accompanied by the threat that decision problems will be confronted which call for more self awareness and computational capacity than the decision maker can muster. Hence, it is pointless, I think, to follow H. A. Simon and other devotees of 'bounded rationality' and seek to weaken the demands of rationality so as to guarantee that prescriptive rationality useful for self criticism coheres with the decision maker's abilities.

We may respond to these problems by acknowledging that principles of prescriptive rationality recommend conforming to their dictates insofar as the agent is capable of doing so while insisting that even when the agent is incapable, it is desirable to develop therapies, educational programs and technologies which enhance his capacity to conform better. We will rest satisfied with our prescriptive norms of rational choice as long as (i) they are applicable in a certain important category of sufficiently computationally undemanding cases free of emotional disturbance and (ii) therapies and technologies for enlarging the domain of applicability are available or are worth developing.

Suppose then that Sam is, indeed, in touch with his beliefs and desires, understands his principles of choice and has enough logical omniscience to identify the option (say, playing the piano) which is admissible according to his principles given his beliefs and desires.

If at the time Sam has figured all this out Sam takes for granted that he will choose rationally (i.e., choose an admissible option), Sam has the omniscience sufficient to conclude that he will choose to play the piano. That is to say, at that time, Sam is certain that he will play the piano and that he will not offer up his wallet.

But if that is Sam's view at the time he applies his principles of choice, from his point of view at that time, surrendering his wallet is not optional for him; for it is inconsistent with his state of full belief that he surrender his wallet – a point he can easily recognize.

There is no contradiction in this result. What it shows is that under the given assumptions Sam has only one option – the uniquely admissible option of playing the piano. Vacuity, not inconsistency, is the trouble. The set of feasible options and the set of admissible options coincide.

The argument illustrated by our example generalizes. If in addition to having the logical omniscience and self knowledge requisite to applying his principles of choice to identify a set of feasible options, the agent is convinced that he will restrict his choice to the admissible options, no inadmissible options are feasible. The principles of choice are applicable; but they are vacuously applicable. They cannot be used to reduce a feasible set to a proper subset of admissible options.

If we are driven to this conclusion, principles of rational choice become useless for the purposes of self policing of decisions.

It will not do to suggest that during the process of deliberation prior to figuring out which options are admissible, Sam has not ruled out the inadmissible options. As long as Sam has not identified the relevant values and beliefs and performed the requisite calculations, the principles of rational choice have not been applied. Only when all the information is in and the calculations have been made has a successful application of the principles of choice to determine admissibility been made. But at that point, the assumption that Sam will choose an admissible option precludes inadmissible options from being feasible. The set of admissible and the set of feasible options coincide. The application becomes vacuous.

Frederic Schick has, in effect, suggested in the face of this predicament that we jettison the idea that principles of rational choice are norms for self criticism.[11] He favors the view that they are principles for the prediction and explanation of choices. In so doing, he acknowledges the main point I mean to press – to wit, that seeing such principles as explanatory and predictive coheres poorly with using them nonvacuously as norms for self criticism.

The alternative view, which I favor, seeks to preserve the status of principles of rationality as nonvacuous norms for self criticism. But if the deliberating agent takes for granted the 'smugness assumption' which asserts that the deliberating agent will choose rationally, the principles of rationality cannot be nonvacuously self applicable. At best, they can be vacuously applicable.

On the other hand, if the smugness assumption is abandoned, the principles of rationality remain self applicable; but now they can be nonvacuous. From the perspective I favor, therefore, the smugness assumption ought to be abandoned.

The argument does not prevent the agent X from predicting that agent Y will choose an admissible option or that X himself will choose ratio-

11 See F. Schick (1979, p. 243). See also Levi (1986, ch. 4, sec. 3) for a further discussion of views on this issue.

nally in some future deliberation. Prediction is precluded only for X predicting his own rational choice in the current context of deliberation. The asymmetry between the first person and third person perspectives implied thereby does not imply a privileged access. It derives from the assumption that canons of rationality are to be used in self criticism in the context of deliberation – that is, in identifying what is to be chosen. I shall return to this point shortly.

Perhaps, it will be argued that even if the decision maker cannot take for granted or be certain that he will choose rationally, he can at least judge that it is probable that he will choose rationally.

But our assumptions about necessary conditions for judgments of feasibility preclude this as well. Indeed, it can be argued that the decision maker cannot make any coherent judgments of credal probability relevant to action concerning what he will choose in the current deliberation.

Consider Sam again. Suppose Sam prefers playing Chopin to yielding his wallet. He is offered a bet as to whether he will play or yield his wallet where he wins $100 if he plays and nothing if he yields his wallet. If he has credal probabilities for what he will choose, the amount he will be willing to pay will be controlled by his credal probabilities. It is clear that Sam should be prepared to pay $100 for the bet. To see this, observe that Sam will clearly prefer playing and taking the bet for any fee less than $100 to playing and refusing to pay that fee and will prefer yielding the wallet and refusing the bet for any fee to yielding the wallet and accepting the bet for a fee. So we need to compare playing and taking the bet with paying and refusing. Given that Sam prefers playing to paying, he should prefer playing and taking the bet to paying and refusing. Moreover he should do so for any fee short of $100. Thus, the 'fair betting rate' for the hypothesis that he will play is 1 and, so it seems, Sam is certain that he will play. Hence, that option and that option alone is feasible for Sam. The admissible and the feasible set coincide.

The upshot is that if Sam is to deploy criteria for choice to determine what he should do, he must not make any judgments as to the probability as to what he will do. A fortiori, he should not make any judgments as to the probability that he will choose rationally. To be an agent crowds out being a predictor.

Suppose that Sam is convinced that he is able to play Chopin through his own choice but is not able to play Chopin through his choice while suffering an asthma attack. I have said that Sam can consistently be convinced of the truth of both claims. Suppose also that Sam is in doubt as to whether he is suffering an asthma attack. Then he must be in doubt as to whether his choice is efficacious and, hence, cannot judge choosing to play to be feasible. (Trying to play may be feasible; but that is a different matter.) We shall assume that Sam is certain that no asthma attack will occur.

These remarks apply, as I have said, to Sam's judgments (or, for that matter, to the judgments of any deliberating agent X) at that stage in deliberation where the agent has identified his values, convictions and options sufficiently to apply the principles of rationality to the evaluation of the admissibility of these options. They do not apply to Sam's evaluation of the rationality of Y's choices or to the rationality of Sam's choices in some other future context of deliberation. Sam can attempt to identify Y's beliefs and values and Y's judgments of feasibility and then apply the principles of rationality to determine how someone in Y's position should choose and still make a prediction as to what Y will choose. If Sam regards Y to be coherent in his judgments, he will not attribute to Y a prediction as to how Y will choose but he can coherently make a prediction of his own.

Since Y can be Sam himself in some future deliberation, we can consider Sam's situation prior to facing the decision whether to play or pay. Suppose Sam is convinced before the Moment of Truth that he will face a decision. There is nothing in our argument which precludes Sam predicting that he will choose an admissible option – i.e., choose rationally. But since he is convinced that he will face a decision where he will choose, Sam is also committed prior to the Moment of Truth to the view that at the Moment of Truth he will cease taking for granted that he will choose rationally.

Moreover, prior to the Moment of Truth, Sam may be in a position to predict what his values and beliefs will be at the Moment of Truth and to determine which of the options he will judge feasible at the Moment of Truth are admissible. Given his conviction prior to the Moment of Truth that he will choose rationally, he will be committed to a prediction as to which act Sam will choose at the Moment of Truth. If there is exactly one admissible option (playing Chopin), the prediction should be that that option will be chosen. If there are several, the prediction will be that one of those admissible options will be chosen.

With the passage of time, Sam arrives at the Moment of Truth. If Sam's prior prediction that Sam will face a decision at that point is borne out, Sam will cease being convinced that he will choose rationally. Moreover, he will cease being convinced as to which of the specific feasible options he will choose.

Thus, Sam will have modified his state of full belief by 'contraction' –

i.e., by giving up some full beliefs. According to the belief-doubt model of inquiry, changes in states of full belief call for justification. What can justify the change in this case?

Observe, at the outset, that if Sam does not give up the prediction that he will choose rationally and, more specifically, that he will play Chopin, then, if he is to preserve coherence, he will have to give up the prediction that he faces a choice between playing and paying at the Moment of Truth. So he will have to give up some prior assumption at the Moment of Truth if coherence is to be preserved. That is coerced by the demands of coherence. The problem of justification concerns what to give up.

A strong case can be made for giving up the prediction that he will choose rationally. Suppose that Sam instead abandons the prediction that he will face a choice between playing and paying. I am supposing that no new information has been obtained by Sam during the interim between his initial predictions and the Moment of Truth which would offer independent reasons for giving up this prediction, so that the only reason for abandoning the prediction that he will face a choice is to preserve coherence and preserve the prediction that he will choose rationally between playing and paying. This strategy is self defeating. Sam can choose rationally between playing and paying only if he faces a choice between these two options. If the only option he faces is playing, he cannot choose rationally between these two options. He may, perhaps, be said to choose in a degenerate sense and, indeed, again in a degenerate sense, to choose rationally. But he is not choosing rationally in the respect in which it was predicted that he would. The upshot is that Sam must abandon the prediction that he will choose rationally between playing and paying whether he retains or abandons the prediction that he will choose between playing and paying. Under the circumstances, abandoning the prediction that he will choose between playing and paying entails a gratuitous reduction in the information available to Sam in his state of full belief. In the absence of an independent justification, he should not do it.

It may, however, be objected that Sam will also have to abandon the prediction that he will play Chopin. Even though Sam cannot claim at the Moment of Truth that he will choose playing over paying, he can at least predict that he will play. The objection is that in retaining the claim that he will face a choice between playing and paying, he abandons this prediction as well. Thus, it may appear that giving up the prediction that Sam faces a decision at the Moment of Truth may not be gratuitous after all.

This objection, however, is not compelling. We are supposing that Sam would not have been convinced that he would play Chopin at the Moment of Truth had he not predicted that he would choose rationally. Since the prediction that he will choose rationally is going to have to be given up anyhow, the prediction that he will play incurs no further loss of explanatory or informational value. Sam is justified in giving up the prediction that he will play.[12]

This would not be true if prior to the Moment of Truth Sam was certain that he would play for other reasons. For example, Sam might have been convinced before the Moment of Truth that he would be incapable of surrendering his wallet because he had already been robbed. But given that information, he should not predict beforehand that he would choose rationally between playing and paying or, indeed, that he would choose between these options at all. And when the Moment of Truth comes, he should not abandon his prediction that he will play unless he has independent good reason for doing so.

When, however, the sole basis for the prediction that he will play is that he will choose between playing and paying rationally (together with the assumptions leading to the conclusion that Sam prefers playing to paying), giving up the rationality assumption warrants giving up the prediction that he will play. And there is no need to justify giving up the prediction that he will choose between playing and paying rationally. Given the other background assumptions, coherence in judgments of feasibility requires doing so at the Moment of Truth.

The cogency of these arguments depends critically on my contention that norms of rational choice should be nonvacuously applicable by the decision maker in policing his deliberations. If one insists on regarding such norms as functioning primarily as principles of an explanatory and predictive theory, the argument fails. The idea that belief-desire models of human behavior controlled, at least in idealization, by principles of rationality constitute an explanatory and predictive theory of human behavior has been widely supported. The arguments I have adduced do not refute this assumption unless the claim that principles of rationality are to be used as norms for self policing is endorsed.

I question the status of principles of rationality as explanatory laws on quite independent grounds. It is well known that we lack the computational capacity, memory and psychic stability to satisfy principles of rationality except in limiting cases. To finesse this difficulty, it is often

12 Appeal is being made here to accounts of contraction which recommend violations of the so-called Recovery Postulate under the conditions envisaged in the text. The Recovery Postulate is discussed and defended in Peter Gärdenfors (1988, ch. 3.4). Violation of this postulate is defended in Levi (1991b, ch. 4, sec. 5).

said that models of rational behavior are 'idealizations' just as theories of ideal gases are, so that they have the kind of explanatory force which is found in the natural sciences. I do not think the analogy is apt. Some real gases approximate the conditions of ideal gases. Human agents never remotely come close to satisfying conditions for ideal rationality if for no other reason than that we lack logical omniscience. To be sure, humans satisfy the conditions well enough in some cases to make decisions meeting the requirements of rationality; but even in those cases, they fail miserably in satisfying all the conditions of rationality. Moreover, modifying the ideal theory so as to improve upon the approximation is always regarded as worthwhile in science. If principles of rationality are intended to be explanatory, we should seek to replace rational explanation by a better theory. Consequently we should abandon the initial principles of rationality as an explanatory theory except perhaps in the way in which discarded scientific theories are used instrumentalistically as crude first approximations. I contend that we should not jettison theories of rationality so easily as that. We know very well that our full beliefs as to what is true are full of logical inconsistencies but we continue or should continue to attach importance to the desirability of removing inconsistencies when we recognize them. Rather than abandoning models of rationality, we should seek instead to devise techniques and therapies which enhance our capacities to do better. In this respect, models of rationality bear a closer resemblance to models of health and mental health. They are, for this reason, normative rather than explanatory and predictive.[13]

Even if this is conceded, however, the conclusions I have been advancing may be resisted. Perhaps models of rationality are designs for better human agents. But we may make such designs without deploying them in self policing. We can design automata to behave in desirable ways without insuring that the automata use the principles of design to police their own behavior. Normativity alone will not bring the conclusions I

13 Akeel Bilgrami (1991) mounts a convincing attack on theories of so-called wide content, direct reference and the like because of their inability to show how content so construed can be deployed in the explanation of human behavior. The weak link in Bilgrami's argument, so it seems to me, is his assumption (shared with many of those he criticizes) that the primary function of appeals to beliefs and desires is in explanation of behavior. However, it seems to me that this reservation with Bilgrami's argument does little to damage it. Even if principles of rationality are norms rather than laws explaining behavior, we should want to claim that were we rational agents completely satisfying the dictates of such principles, self criticism would be otiose, rational angels would be rational automata and our behavior would be explainable by these principles. Indeed, were this not the case, we would regard the principles of rationality as somehow defective as norms for use in self criticism.

have been defending. Appeal must be made to the use of the norms in self policing of deliberation. Thus, those who resist the conclusions I am advancing must insist on external policing to the exclusion of internal policing.

The point is well taken. External policing can take the form of deploying norms of rationality as blueprints for rational automata. There is no clash between using principles of rationality for explanatory and predictive purposes, on the one hand, and using them prescriptively for designing rationally acceptable conduct. Tension arises only if, in addition to using them for external policing, one seeks to use them for internal policing. In that event, the blueprints can no longer be for rational *automata.* Agents will satisfy the requirements for rational health only if they apply the principles of choice to evaluate their options. But, in that case, neither they nor we, the outside agents, can regard them as predicting their own choices. Rational automata can predict their own choices. Rational agents cannot.

The tedious mental gymnastics of the paragraphs dealing with Sam before and at the Moment of Truth are intended to undermine the impression one might receive that the asymmetry between the first personal point of view and the third personal point of view smacks of mystery or irrationality. Once the roots of the asymmetry in the demand that principles of rational choice be nonvacuously self applicable are recognized, there should be neither mystery nor irrationality.

Mysterious or not, the implications of the approach I am pressing are far reaching for accounts of rational decision making, rational probability judgment and rational value judgment. I have discussed some of these ramifications elsewhere. Space does not permit elaboration on them here.[14]

I declared my interest in the accounts of rational choice as deriving from my interest in a systematic development of the pragmatist belief-doubt model of inquiry as a prescriptive account. On the basis of this interest, I defended the idea that principles of rationality should be relatively weak, context independent principles, applicable in a wide variety of contexts and that they should be relevant principles for assessing both wise choices and apt feelings.

Is there any element in the pragmatist approach which might argue for or against insisting on construing the prescriptions of rationality as nonvacuously applicable in self criticism and deliberation rather than as recipes for designing rational automata? I have not been able to find any

14 See Levi (1987, pp. 193–211, 1991c, ch. 4, and 1992, pp. 1–20).

clear indication in the writings of the classical pragmatists that they recognized the need to consider the issue. Authors who in one way or another have intimated their pragmatist sympathies subsequently have felt free to adopt positions which, if the argument of this essay is right, abandon the use of principles of rationality for purposes of self criticism.

Consequently, my appeal to an interest in articulating a pragmatist account of problem solving inquiry as a means for identifying a conception of rationality has exhausted its resources without fully settling the question. If one appeals to Dewey's famous interest in education aimed at training students in the methods of well conducted problem solving inquiry, one will still face the issue of deciding whether the training will require inculcating a capacity for self criticism or whether it is enough to produce well trained seals. I have no doubt that Dewey would have opted for the former over the latter alternative. But appeals to authority will not settle the issue. Still if there is any remnant of the Kantian view of autonomy worth preserving from a pragmatist perspective, it is to be found in the nonvacuous applicability of standards of rationality in self criticism.

Opponents and proponents of preserving this remnant should recognize, however, the presence of the tension between the explanatory use of principles of rationality and their use as norms for self criticism. The tension is severe and fraught with consequences deserving serious philosophical reflection.

REFERENCES

Bilgrami, A. (1991), *Belief and Meaning,* Oxford: Blackwell.

Dewey, J. (1938), *Logic: The Theory of Inquiry,* New York: Holt.

(1941), "Propositions, Assertibility and Truth," reprinted in *Dewey and His Critics,* ed. by S. Morgenbesser, New York: Hackett, 1977, 265–82.

Dewey, J. and Tufts, J. H. (1932), *Ethics,* New York: Holt.

Gärdenfors, P. (1988), *Knowledge in Flux,* Cambridge, MA.: MIT Press.

Gibbard, A. (1990), *Wise Choices, Apt Feelings,* Cambridge, MA.: Harvard University Press.

Levi, I. (1980), *The Enterprise of Knowledge,* Cambridge, MA.: MIT Press, ch. 4.

(1983), "Doubt, Context and Inquiry," in *How Many Questions,* ed. by L. Cauman et al., New York: Hackett, 25–34.

(1986), *Hard Choices,* Cambridge: Cambridge University Press.

(1987), "The Demons of Decision," *The Monist 70,* 193–211.

(1991a), "Chance," in *Philosophical Topics: Philosophy of Science,* ed. by L. Nissen, vol. 18, New York: Hackett, 95–121.

(1991b), *The Fixation of Belief and Its Undoing,* Cambridge: Cambridge University Press.
(1991c), "Consequentialism and Sequential Choice," in *Foundations of Decision Theory,* ed. by M. Bacharach and S. Hurley, Oxford: Blackwell, ch. 4.
(1992), "Feasibility," in *Knowledge, Belief and Strategic Interaction,* ed. by C. Bicchieri and M. L. Dalla Chiara, Cambridge: Cambridge University Press, 1–20.
Peirce, C. S. (1984–86), in *The Writings of Charles S. Peirce,* ed. by M. Fisch et al., Indianapolis: Indiana University Press, vol. 2, 98–102, and vol. 3, 300–301.
Schick, F. (1979), "Self Knowledge, Uncertainty and Choice," *British Journal for the Philosophy of Science 30,* 235–52.

3
The logic of full belief

1

That logic has a prescriptive use as a system of standards regulating the way we ought to think is fairly noncontroversial. Frege (1967, pp. 12–13), the resolute antipsychologist, did not deny it. But Frege also thought that logic was a system of truths. The prescriptive force of logic derived from this together with a fundamental value commitment to believe what is true. According to this view, we ought also to believe the true laws of physics and to reason in conformity with them. To be sure, logic is distinguished from physics in virtue of the special status of its truths as "the most general laws, which prescribe universally the way in which one ought to think if one is to think at all" (Frege, 1967, p. 12). Still the prescriptive use of logic is not what distinguishes it from physics. Both logic and physics as human activities aim to ascertain truths regarding certain subject matters. Logic differs from other natural sciences only insofar as its subject matter is objective truth itself. The imperative to seek truth and reason in conformity with it ought to be obeyed regardless of whether the true laws invoked are laws of objective truth as Frege took them to be or laws of physics.

Ramsey (1990, p. 80) rightly pointed out that undertaking to fulfill a value commitment to believe what is true is jousting at windmills.

> We may agree that in some sense it is the business of logic to tell us what we ought to think; but the interpretation of this statement raises considerable difficulties. It may be said that we ought to think what is true, but in that sense we are told what to think by the whole of science and not merely by logic. Nor, in this sense, can any justification be found for partial belief; the ideally best thing is that we should have beliefs of degree 1 in all true propositions and beliefs of 0 degree in all false propositions. But this is too high a standard to expect of mortal men, and we must agree that some degree of doubt or even of error may be humanly speaking justified.

For my good friend and admired colleague, Charles Parsons. Thanks are due to Horacio Arló Costa for good advice.

Even if we grant that in some sense or other the laws of logic are conceptual necessities, they are not conceptual necessities in a sense that guarantees that we are capable of assenting to them in either language or behavior upon demand. In this respect, too, they are like the true laws of physics.

In one respect, however, they may be understood to be different. To the extent to which we fail to fulfill the commitment to believe logical truths, the remedy for the failure is to be found in more training (in logic and mathematics and their use in applications), in the use of prosthetic devices (such as paper and pencil, the use of tables and computer technology), and in various forms of psychotherapy. When we fail to fulfill the commitment to believe extralogical truths, we must supplement the use of these therapeutic devices with *inquiry.* We consult the testimony of the senses and reliable witnesses and authority and we draw ampliative inferences as in statistical and inductive inference or in choosing which rival theories to believe.

The laws of deductive logic are, I am suggesting, different from the laws of physics not only because they are true but because rational agents are committed by the standards of rational health to full belief that they are true. We are not committed at any given time or period of time t to fully believe (or to be certain of) at that time t all truths; but we are committed at that time t to have full beliefs that are logically consistent and to fully believe all the logical consequences of what we fully believe at that time. We no doubt fail to fulfill this commitment and, indeed, cannot do so in forming our doxastic dispositions and in manifesting them linguistically and in our other behavior. We are, nonetheless, committed to doing so in the sense that we are obliged to fulfill the commitment insofar as we are able to do so when the demand arises and, in addition, have an obligation to improve our capacities by training, therapy or the use of prosthetic devices provided that the opportunity is available and the costs are not prohibitive (Levi, 1991, ch. 2).

Thus, we are committed to fully believing at every time t the truths of logic because we are committed as rational agents to having full beliefs at t that are at once consistent and closed under deductive consequence. Insofar as we fail to fulfill this commitment, the defect is in us and we must seek remedies for our deficiencies. Failure to believe the true laws of physics reflects no such defect in our doxastic condition. Indeed, it is often a mark of a healthy mind to be able to acknowledge our ignorance. Removing such ignorance calls for a *change* in doxastic commitment – that is to say, a change from a commitment to one system of consistent and closed full beliefs to another such commitment. This view stands in opposition to Frege's (and to Russell's) according to which our commitment to believing the truths of logic derives from our commitment to believing truth including in particular the true laws of the science of truth. According to this view, all doxastic changes alter the extent to which we fulfill our single commitment to believe what is true. The commitment never changes. Only our attempts to fulfill it do.

The opposition might not seem as sharp as I am advertising it to be, given a certain explication of the special status of logical truth. According to the standard of doxastic health, rational agents are committed to fully believing all logical theses and, hence, to judging (i.e., to being certain) that theses of deductive logic are true. In this sense, a logical truth is any candidate belief all rational agents are committed to fully believing according to the standards of conceptual health.

The opposition may be mitigated but it is not eliminated in this way. As I understand Frege's view, the status of logical truths as truths does not derive from standards of rational doxastic health and does not characterize our doxastic commitments. We may be committed to believing logical truths, but our commitment is the same with respect to the laws of physics. Frege does not allow for a distinction between failures of commitment that are removed by therapy, training and the use of prosthetic devices on the one hand and failures that are removed by inquiry on the other.

The contrast I charge Frege with neglecting or denying can be seen as a contrast between two senses of change of belief. Some changes of belief are best seen as attempts to fulfill unfulfilled commitments. These are of the sort that Peirce described as "explicative". Deductive inference is the classic example. According to Ramsey, who endorsed Peirce's distinction, deduction "is merely a method of arranging our knowledge and eliminating inconsistencies or contradictions" (1990, p. 82). I prefer to understand such changes over time as changes in belief that incur no change in doxastic commitment during the process but only a change in fulfilling such commitments. Ampliative reasoning, by way of contrast, seeks to justify changing from a relatively uninformative commitment to full belief to a stronger one. Ramsey, following Peirce, saw ampliative reasoning as a way of acquiring new knowledge. For Ramsey, the standard ways of doing so are through observation, memory and induction. I would delete memory from this list or at least accord it an ambiguous status since the use of memory is often in the service of fulfilling commitments already undertaken. But setting aside the ambiguous status of memory, my proposed gloss on the Peirce-Ramsey contrast seeks to draw the distinction between explicative and ampliative changes in belief in terms of changes that are improvements in the fulfillment of commitments already undertaken or are the product of improvements in capacity to fulfill such commitments and changes in commitments that are the product of inquiry.

Elaboration of this view of the matter calls for three main lines of investigation. First, one must provide some sort of account of what a state of doxastic commitment (commitment to full belief) is. In terms of such an account, one can then discuss failures to fulfill doxastic commitments and changes in belief that represent improvements in fulfilling commitments. One also can provide an account of changes in doxastic commitment. Second, one must provide some sort of account of the techniques involved in improving fulfillment of commitment. Armchair philosophical reflection will not be sufficient here. Investigation of the psychology of learning, the development of appropriate psychotherapies, the identification of relevant technologies and the study of logic and mathematics are all important. Finally, one needs an account of the intelligent conduct of inquiry. In this connection, philosophical reflection on the invariant features of intelligently conducted inquiry can play some role provided it is recognized that the features of inquiry proposed should be made to square somehow with what is noncontroversially recognized as the intelligent conduct of inquiry.

Ramsey proposed to distinguish between a lesser logic or the logic of consistency and a larger logic or the logic of discovery or induction (Ramsey, p. 82). As I am glossing him, the logic of consistency for full beliefs is logic in the sense in which logic characterizes the doxastic commitments undertaken in a given state of full belief. Given any agent X at any given time *t*, the logic of consistency specifies conditions that X's full beliefs at *t* should meet to satisfy the commitments X has undertaken at *t*. This logic of consistency is, therefore, addressed to partially answering the first of the three questions just itemized: What are the commitments generated by a state of full belief?

Ramsey conceded that insofar as we are in a position to characterize the set of logical truths independently of their role in characterizing doxastic commitments, we might proceed as Frege had done by first specifying the set of logical truths as "objective" truths and then prescribing a commitment on the part of rational agents to those truths and thereby obtaining the logic of consistency. Ramsey registered an unexplained reservation with this concession endorsed by him for the sake of the argument. One possible reason is that insofar as logical theses can be sensibly explicated without appeal to their role in characterizing doxastic commitments, logical truths are characterized without any indication of their intended application. They constitute a formal structure that may and does have many applications including the prescriptive one under consideration here.[1] As a constituent of a formal structure, a logical thesis is not a truth at all. Frege's claim may then be understood as claiming that one application of a pure formal logic is to the systematic characterization of the objective logical truths. Skeptics might reasonably wonder whether Frege succeeded in furnishing an application of pure logic at all.

1 Logic as understood by Koslow (1992) studies diverse "implication structures" and their properties. Deductive logics (either propositional or first order quantificational logic) are implication structures of certain kinds. I am considering cases where the implication structures are atomless Boolean algebras of potential states of belief closed under meets and joins of sets of states of any cardinality (Levi, 1991, ch. 2). Here the "implication relation" is the binary relation partially ordering the potential states in the algebra with respect to how well potential states relieve doubt. Using techniques described by Koslow (1992, ch. 20), the given implication structure for potential belief states can be extended so that quantification can be introduced while continuing to avoid attributing syntactical structure to the potential belief states. The additional elements belonging to the extended implication structures are neither additional states of belief nor constituents of states of belief.

Belief states belonging to subalgebras of the set of the algebra of potential states of full belief may be represented by sentences in a regimented language $\underline{L}$ where the logical consequence relation between sentences or sets of sentences in $\underline{L}$ may be understood as generating an implication structure that represents the subalgebra of potential states of full belief by preserving the consequence relation between belief states. Hence, the implication structures at the focus of attention when we think of logic of full belief are algebras of potential belief states and secondarily the structures determined by the relation of logical consequence for the language $\underline{L}$ used to characterize such structures. Frege apparently was interested in implication structures where the domain of entities is composed of thoughts distinct both from potential states of full belief and from linguistic entities. I have no understanding of the applications intended when thoughts are introduced.

Nothing in this paper is intended to question the importance of the study of implication structures in general. I am focusing on structures relevant to the understanding of logic as characterizing standards of doxastic health – the logic of consistency for full belief – and its relation to the study of implication structures characterizing logics of truth.

Hintikka (1969, p. 5) writes: "A branch of logic, say epistemic logic, is best viewed as an *explanatory model* in terms of which certain aspects of the workings of our ordinary language can be understood". A logic so conceived need not even be an implication structure and rightly so when we consider logics of probability judgment, preference etc. But even in the case of logics of knowledge and belief where Hintikka and I might agree that the study of implication structures is central, we seem to differ regarding the role of logic in explaining "the workings of our ordinary language". For me, unlike Hintikka, the primary focus is on the critical control of the attitudes in general and of belief in particular. I use language in offering a systematic account of the standards for doxastic health I favor and acknowledge, as one must, that linguistic means are used to communicate beliefs and other attitudes. I do not deny that the regimented languages I use for systematic discussion are implication structures of a certain kind. But their logics are used for the purpose of examining the doxastic commitments of rational agents. Explanation of "the workings of language" are of, at best, marginal interest and relevance to this project.

Waiving that objection (which may not have been Ramsey's worry), when it comes to deductive logic, there is a *prima facie* case for saying that the "logic of consistency" coincides with the "logic of truth" - that is, a science of logical truths.

Ramsey's chief concern was not with the details of probability logic - that is, a logic regulating degrees of belief in the sense of degrees of subjective or credal probability. He argued that there is a logic of consistency for judgments of credal probability just as there is for full beliefs. If an agent judges hypothesis h to be probable to degree $\frac{1}{3}$ the agent is, thereby, committed to judging $\sim h$ to be probable to degree $\frac{2}{3}$ just as the agent who fully believes that h is committed to full belief that $h \vee f$. But the commitment to credal probability judgment involves no commitment to judgments that are true or false except those required by the logic of full belief (1990, p. 93). Probability logic requires the agent to fully believe and, hence, assign credal probability 1 to all truths of deductive logic. If the agent fully believes some extralogical propositions, the agent is committed to judging them probable to degree 1 and to fully believing and assigning probability 1 to their logical consequences. These are the only truths to which probability logic requires a commitment. For this reason, Ramsey denied that the logic of consistency for probability coincided with a logic of truth for probability. Ramsey explored a logic of truth for probability appealing to reasoning about frequencies. But as Ramsey appreciated, the relation of frequencies to subjective probability provides no basis for equating the logic of consistency and of truth for probabilities.

I wish to press Ramsey's point still further and to argue that his concession to Frege's view in the case of full belief needs to be modified. The logic of consistency for full belief, so I will argue, does not coincide with the logic of truth.

2

Why should we insist on requiring deductive closure and consistency as part of the standard of doxastic health? What is the point of insisting that rational agents are committed to meeting such a standard?

Decision theorists, economists, statisticians and philosophers interested in rational deliberation and inquiry commonly presuppose that agents who assign positive credal probability to a hypothesis judge that that hypothesis is possibly true. Presumably the linguistic and behavioral manifestations of the judgment of possibility, like such manifestations of judgments of credal probability, reveal propositional attitudes that are subjective in the banal sense that they are propositional attitudes. I contend that commitment to judgments of subjective possibility and impossibility is tantamount to a commitment to a set of full beliefs. Because of the central relevance of judgments of serious possibility to deliberation and inquiry, I call such judgments, judgments of *serious possibility.* The *standard for serious possibility condition* (SSP) formally characterizes the bridge between commitments to full belief at a time and judgments of serious possibility at that time.

(*SSP*): For every X and *t,* X is committed at *t* to fully believe that *h* if and only if X is committed at *t* to judge ~*h* impossible. For every X and *t,* X is committed not to believe that *h* at *t* if and only if X is committed at *t* to judge that ~*h* is possible.

According to the SSP principle, an agent at a time may have commitments to fully believe and commitments not to fully believe. SSP by itself does not rule out cases where X lacks both a commitment at *t* to believe that *h* and a commitment at *t* not to believe that *h.* If such a lack of commitment were to arise, X would have no commitment to judge *h* possible or to judge *h* not possible. X's state of full belief would fail to serve the function of a standard for serious possibility with respect to *h.*

It would, of course, be undesirable to rule out cases where the standard for serious possibility does not assess serious possibility with respect to *h.* SSP states only that if assessing *h* with respect to serious possibility is off the charts, so is commitment to believe that *h* and commitment not to believe that *h.*

However, when attention is restricted to the domain of objects that are candidates for doxastic commitment and modal judgments, SSP is too weak and needs supplementation by the following *principle of disbelief* (~B):

(*~B*): X is not committed at *t* to full belief that *h* if and only if X is committed at *t* not to believe that *h.*

SSP and ~B alone do not prevent X at *t* from being committed to judge that both *h* and ~*h* are impossible. Supplementing SSP with the principle of deductive consistency does. But consistency and SSP do not prohibit X at *t* from being committed to judging ~*h* and judging $h \wedge {\sim}f$ both impossible while judging both *f* and ~*f* possible. The addition of ~B and intuitionistic deductive closure prevents this. But unless we insist on closure under classical deductive consequence, there is no way that X could be committed to judging *h* to be possible (impossible). If X is committed to (not) fully believing that ~*h,* X would be committed to judging $\sim\sim h$ impossible (possible) without commitment to judging *h* impossible (possible). Hence, X could not be committed to suspending judgment concerning *h* and $\sim h$; for this requires judging both *h* and $\sim h$ to be possibilities. We could add a clause to SSP providing for judging *h* impossible (possible) if and only if $\sim h$ is (not) fully believed. But adding this clause would ensure closure under classical logical consequence. The principles of classical deductive consistency and closure are needed as supplements to SSP and ~B to ensure that judgments of serious possibility can provide a basis for making coherent credal probability judgments in a self critical way.[2] As I shall argue shortly, another principle is needed as well. Before turning to that matter, I shall elaborate some more on the difference between commitment to full belief and fulfilling such commitments.[3]

3

The previous discussion suggests that the sentence "X fully believes at *t* that *h*" may be understood in three ways: It may represent a feature of X's doxastic commitment at *t*, it may represent a set of dispositions to assent and to act that generate such a doxastic commitment or fulfillment of the doxastic commitment already undertaken or it may represent

2 There may be contexts where intuitionist logics (implication structures) gain useful application; but not as the logic of consistency for full belief. Closure under a classical deductive consequence relation should be required when states of commitment to full belief are understood to be states of commitment to standards for judging serious possibility. Another version of the case against an intuitionistic understanding of deductive closure is based on the arguments advanced in Levi (1991, ch. 2) for requiring that the set of potential or candidate belief states have a structure that is at a minimum that of a boolean algebra rather than of a pseudo boolean algebra. My argument in brief is that if x and y are potential belief states, there ought to be a potential belief state that is a state of doubt or suspense with respect to x and y - that is, the join of x and y. Those who insist that the structure is that of a pseudo boolean algebra say that a potential belief state x may have a "pseudo complement" $\neg x$ such that join of x and $\neg x$ is strictly stronger than the maximally skeptical belief state of being committed to judging only logical truths to be true. Yet, there is no potential belief state relative to which one could judge the join of x and $\neg x$ impossible. One could not fully believe it to be false. Prohibiting such potential belief states constitutes a dogmatic foreclosing of inquiry and ought to be rejected. The set of potential belief states ought to constitute at a minimum a boolean algebra reflecting the demands of a classical two valued logic.

3 One could argue that satisfying the strict standard of commitment characterized by deductive consistency and closure, SSP and ~B is really unnecessary. The extent of logical omniscience demanded of rational agents need only suffice to meet the demands of the problem being addressed. One can get by with something substantially short of full omniscience and still come out with the same conclusions one would have reached were one fully omniscient. That is often true. Still there is no upper bound on the complexity of the problems we may face. The commitments enjoined by the strict standard are an acknowledgment of that point.

manifestations of such dispositions. I shall follow the policy of expressing the first sense by "X is committed at *t* to full belief that *h*", the second sense by "X fully believes at *t* that *h*", and the third sense by "X manifests at *t* full belief that *h*".[4]

The sentence *h* that appears in the *that* clauses is to be understood here as part of a linguistic apparatus used to represent X's state of full belief or doxastic commitment. It need not be ingredient in the language used by X in manifesting X's full beliefs. It does not matter whether *h* is a sentence in some natural language or in some suitably regimented language $\underline{L}$. The sentences in $\underline{L}$ can be used to represent X's doxastic commitments, the extent of X's fulfillment of these commitments by having the requisite full beliefs and, indeed, the manifestations of such commitments. The sentences and sets of sentences are not used to manifest beliefs as one does in sincerely making a statement. They are used instead to represent doxastic commitments just as a geometrical structure is used as a phase space to represent mechanical states.

Thus, I propose to represent X's state of full belief (insofar as it is representable in $\underline{L}$) by a set $\underline{K}_{X,t}$ of sentences in $\underline{L}$ that I shall call X's *corpus* at *t*. X is committed at *t* to full belief that *h* if and only if *h* is a member of $\underline{K}_{X,t}$. I employ a regimented language $\underline{L}$ that comes with a consequence relation $\underline{A} \vdash h$ between sets of sentences in $\underline{L}$ and a sentence in *h* in $\underline{L}$ for a sentential or first or deductive logic. The logical theses of $\underline{L}$ are consequences in $\underline{L}$ of any set of sentences in $\underline{L}$. I shall use sets of sentences closed under deductive consequence and consistent sets of sentences. These products of the structure $\underline{L}$ are introduced without any consideration of truth or falsity.

X believes at *t* that *h* if and only if X at *t* has dispositions directly generating the commitment to full belief that *h*. Two features of this characterization need to be explained at least briefly.

First, having the dispositions is not sufficient for full belief. Just as, on some occasions, a person may sign on the dotted line without contracting to do something, so too X might have and manifest dispositions to assent and behave that in some contexts directly generate commitment to full belief that *h* but fail to do so in the given situation. Whether having and manifesting certain dispositions fulfill or fail to fulfill the conditions for directly generating commitment to full belief that *h* cannot

4 In Levi (1991) and Chap. 1 of this volume, I reserved "X fully believes at that *h*" for full belief in the commitment sense and "X fully recognizes at *t* that *h*" for full belief in the dispositional (and sometimes manifestation) sense. For the purposes of this discussion, I wish to focus attention on the consistency or coherence of beliefs understood as attempts at partial fulfillments of doxastic commitments. Hence, the change in terminology.

be settled without appealing to normative considerations. Settling the issue is, however, to decide whether the dispositions in question are interpretable as believing at *t* that *h*. The task of interpretation is to identify the obligations the agent has incurred in acquiring or having those dispositions just as it is when signing on the dotted line is interpreted as making a promise.

The second feature of my understanding of "X believes at *t* that *h*" is that it claims that the X's dispositions *directly* generate doxastic commitments. The reason for insisting on the qualification that the commitments be incurred directly is that X might be committed at *t* to full belief that $h \vee f$ in virtue of fully believing that *h* without fulfilling that commitment by fully believing that $h \vee f$. X has dispositions at *t* that incur a commitment at that time to believe that *h* and, as a consequence, a commitment to believe that $h \vee f$. He may not then have the dispositions to assent and act that would have generated the commitment to full belief that $h \vee f$ whether or not he fully believed that *h* (or that *f*) or was committed to doing so. Thus, the dispositions that directly generate the commitment to full belief that $h \vee f$ partially fulfill the commitment undertaken by X to believe that *h* without generating the latter commitment either directly or indirectly. The very same dispositions directly generating commitment to full belief that $h \vee f$ simultaneously partially fulfill that commitment. Consequently, the notion of direct generation of a doxastic commitment may be replaced in the formula characterizing full belief as follows:

X fully believes at *t* that *h* if and only if X has dispositions to assent and behave at time *t* that not only commit X at time *t* to fully believe that *h* but partially fulfill the commitment.

I take for granted here that commitment to full belief that *h* is also commitment to full belief that $h \vee f$. In so doing, I invoke some normative principles of the logic of consistency for full belief. Without a logic of consistency for full belief (and judgments of serious possibility), there is no system of normative principles characterizing commitment to full belief or the notion of full belief as a set of dispositions to assent and act directly generating such commitments. We return, therefore, to the discussion of this logic of consistency.

4

X's state of doxastic commitment (representable in $\underline{L}$) at *t* shall be represented by the corpus $\underline{K}_{X,t}$ of sentences in $\underline{L}$. *h* in $\underline{L}$ belongs to $\underline{K}_{X,t}$ if and only X is committed at *t* to full belief that *h* (whether X has dispositions directly generating such commitment or not). By the ~B principle, sentence g in $\underline{L}$ but not in $\underline{K}_{X,t}$ represents X's commitment at t not to fully believe that g. SSP commits X at t to judge h seriously possible if and only if $\sim h$ is not in $\underline{K}_{X,t}$. The principle of deductive consistency and closure requires $\underline{K}_{X,t}$ to be deductively consistent and closed. X's full beliefs at t are "fully coherent" just in case X's set of full beliefs at t are representable by the same set of sentences that represents his doxastic commitments at t - that is, X's corpus at t. The logic of consistency for full belief is the set of normative principles spelling out those conditions that a set of sentences in $\underline{L}$ should satisfy to represent X's state of doxastic commitment - that is, X's state of full belief. In that sense they characterize potential states of full belief in terms of potential corpora.

The principle of deductive consistency and closure commits every rational agent X at all times (1) to full belief that $\underline{T}$ where $\underline{T}$ is a logical thesis, (2) to full belief that h if X is committed to full belief that x for every x in $\underline{A}$ such that $\underline{A} \vdash h$. By principle ~B, X is committed at all times (3) not to believe any sentence $\underline{F}$ that is the negation of a logical thesis. These requirements together with SSP commit X at t (4) to judge seriously possible all and only those sentences in $\underline{L}$ that are logically consistent with $\underline{K}_{X,t}$.

Notice that all rational agents at all times are committed to full beliefs representable by logical theses in $\underline{L}$. In this sense, logical theses are universally valid. Are they true? More importantly are the full beliefs they represent true? It is noncontroversial that if X at t fully believes that h, X fully believes that h is true. That is to say, believing that h is judging that h is true. So every agent is committed to judging that logical theses are true. So rational agents are committed to a consensus that theses of deductive logic are true. To the extent that an answer to the question manifests a judgment (full belief) by a rational agent fulfilling his or her doxastic commitment, the answer by universal commitment is: Yes! Logical theses in $\underline{L}$ and the beliefs they represent are true.

Similarly, by universal commitment, logical consequence is truth preserving and logical contradictions are false.

Thus, the theses of deductive logic in $\underline{L}$ represent beliefs to which every agent at all times is committed if their beliefs satisfy the requirements of coherence specified by the principles of deductive consistency and closure. They also characterize a logic of truth by unanimous commitment. If the principles of deductive consistency and closure constituted a complete logic of consistency for full belief, there would be an important sense in which the logic of consistency for full belief coincides with a logic of truth. But a logic of truth so conceived is a logic of truth by unanimous commitment. The logic of truth by unanimous commitment presupposes the normative principles of the logic of consistency just laid down. Whether this characterization of a logic of truth for full belief would have been sufficiently nonpsychologistic to satisfy Frege is a matter I leave to Frege scholars. It is for me. Even so, I do not think the logic of consistency for full belief does coincide with a logic of truth for full belief; for the principle of deductive consistency and closure does not exhaust the logic of consistency for full belief, and the additional principles do not impose commitment to full beliefs representable by additional sentences that are universally valid and, hence, true by universal commitment.

Given a universally mandatory commitment to a deductively consistent and closed system of full beliefs and to ~B, it follows that rational agents are committed to satisfying the following *Opinionation Condition* (OC):

For every X at every time and for any *h*, either X is committed at *t* to full belief that *h*, to full belief that ~*h* or suspension of judgment between *h* and ~*h*.

For every *h* in L, either X is committed to full belief that *h* or X is not so committed. By ~B, it follows that either X is committed to full belief that *h* or committed not to believe that *h*. Consistency rules out being committed to full belief that *h* and at the same time to ~*h*. Commitment to full belief that *h* generates a commitment not to believe that ~*h*. But being committed not to believe that *h* and also not to believe that ~*h* is not ruled out. Even so, deductive closure requires commitment to full belief that *h* ∨ ~*h*. OC thus follows from ~B, deductive consistency and closure when commitment to suspension of judgment between *h* and ~*h* is understood to be a commitment to full belief that *h* ∨ ~*h* without commitment to full belief that *h* and without commitment to full belief that ~*h*. Commitment to suspension of judgment between *h* and ~*h* becomes, via the SSP and ~B principles, equivalent to a commitment to judge both *h* and ~*h* seriously possible.

I take for granted that X's belief at *t* that *h* and, indeed, X's commitment to belief at *t* that *h* may coherently be judged (by X and Y at some time or other) true or false. I also take it for granted that the full belief of some one or other at some time or other that X at *t* judges it possible that *h* may be judged by someone at some time to be true or false. But it would be incoherent of any one at any time to judge X's judgment of serious possibility that *h* at *t* to be true or false. Full beliefs may be judged true or false but not judgments of serious possibility. In judging both *h* and ~*h* to be serious possibilities at *t*, X is no doubt committed to full belief that *h* ∨ ~*h* and by deductive consistency and closure is committed to fully believing it. So this full belief is truth valued and indeed by unanimous commitment is judged true. But X is also committed not to believe that *h* and not to believe that ~*h*. Commitments not to believe are not themselves truth valued. Consequently, insofar as X's state of doxastic commitment at *t* serves as X's standard for serious possibility via SSP and ~B, the logic of consistency of full belief is simultaneously a logic of consistency for judgments of serious possibility. Possibility judgments cannot carry truth values without implying that probability judgments do so as well. Ramsey rightly concluded that the logic of consistency for credal probability judgment cannot coincide with whatever may pass as a logic of truth for credal probability. A parallel observation is appropriate regarding the logic of consistency for judgments of serious possibility and a logic of truth for such judgments. But the logic of consistency for judgments of serious possibility is automatically a logic of consistency for full belief. The logic of truth for full belief (if there be such) cannot coincide with the logic of consistency.

It seems to me, however, that this is not the end of the story. Granted that the logic of consistency for full belief combined with SSP and ~B cannot be a logic of truth for judgments of serious possibility, the logic of consistency for full belief may still remain a logic of truth for full belief. This claim needs some examination.

5

Rational agents are not merely agents who have doxastic commitments that they fulfill to some partial extent and fail to fulfill otherwise. Nor are the prescriptions characterizing doxastic commitments designed merely to be used by other agents to determine the extent to which the agent under study is or is not fulfilling his or her commitments. Rational agents are supposed to use the logic of consistency to criticize their own performance. They are not merely rational automata who may or may not be well designed to conform to the demands of the logic of consistency but are self critical agents who can apply the logic of consistency to their own point of view. Self reflection, so I argue, calls for a logic of consistency that includes deductive logic but calls for a larger network of commitments to full belief than those demanded by deductive logic. It is in this setting that it becomes apparent that the logic of consistency for full beliefs cannot coincide with a logic of truth (if there be such) for full belief.

Self reflecting agents are committed to deductively closed and consistent sets of full beliefs but are also committed to identifying what those states of commitment are. In particular, all agents at all times are committed to beliefs in conformity with the following principle:

(*BB*): If agent X at *t* is committed to fully believing that *h*, X at *t* is committed to fully believing that X at *t* fully believes that *h*.

Since fully believing that *h* presupposes a commitment to fully believe that *h*, fulfilling the commitments generated by this BB-principle requires that an agent who fully believes that *h* fully believe that he or she fully believes that *h*. If X fully believes that *h* but fails to fully believe that X fully believes that *h*, X has not completely fulfilled the commitments undertaken in fully believing that *h* just as X has not succeeded in fulfilling such commitments in failing to fully believe that $h \vee g$. That is to say X has dispositions to assent interpretable as directly generating X's commitment to full belief that *h* but does not have such dispositions interpretable as directly generating commitment to full belief that $h \vee g$.[5]

5 Recall that for the purposes of this discussion, "X fully believes at *t* that *h*" is to be understood to assert that X at *t* has dispositions to assent and behave that directly generate a commitment at *t* to full belief that *h*. X may be committed at *t* to full belief that *h* without fully believing at *t* that *h* - i.e., having the requisite direct commitment generating dispositions to assent and behave. X may, indeed, have the dispositions that directly generate a commitment to full belief that $\sim h$ or, indeed, that directly generate a commitment to suspense between *h* and $\sim h$. In such cases, X is failing to live up to X's commitments to full belief. X's full beliefs are in this sense inconsistent. As a consequence, at least one of the system of normally direct commitment generating conditions must be considered to have failed to be commitment generating at all. For, the direct commitment generating conditions for commitment to full belief that *h* and the corresponding conditions for commitment to full belief that $\sim h$ cannot both be direct commitment generating conditions when the two sets of conditions are jointly satisfied. It becomes a problem for interpretation to determine which of the two sets of conditions, if either, is to be interpreted as commitment generating in the given context. Even so, we may claim that X fully believes that *h* and also fully believes that $\sim h$ in the sense that at the same time X satisfies both the conditions that *would* be directly commitment generating for full belief that *h* if no failure to have commitments required by the logic of consistency were thereby mandated and the corresponding conditions for full belief that $\sim h$.

By the same token, X may have the dispositions at *t* directly generating a commitment to full belief that *h*, but may lack at *t* the dispositions directly generating a commitment to full belief that X fully believes that *h*. In this case too, a failure to fulfill commitments arises that warrants judging X's full beliefs at *t* as inconsistent. We are not to conclude that his doxastic commitments violate the logic of consistency but only that *prima facie* direct commitment generating conditions fail to be commitment generating at all. A problem for interpreting the agent's dispositions to assent and to behave arises.

The dispositions to assent and to other behavior exhibited by X at *t* may be sufficiently confused at a given time that neither X nor anyone else can identify a doxastic commitment that X has undertaken. X cannot be understood to believe that *h*, believe that $\sim h$ or to be in suspense regarding *h* and $\sim h$. Here, too, X has failed to fulfill commitments and X is doxastically inconsistent. The OC condition has been violated.

An automaton might be designed to have dispositions to assent and behave that *we* interpret as simulating manifestations of a doxastic state that is consistent and deductively closed. But we should not understand the automaton as having commitments to full belief that it either fully or partially fulfills (as having full belief in the strict or weaker sense) unless we also are prepared to interpret the automaton as being self critical – perhaps, in virtue of its exhibiting manifestations of dispositions to assent and behave in ways that we can interpret as detecting and removing inconsistencies, failures of deductive closure and eliminating self deception. If an automaton is interpretable as a self critical rational agent, then we interpret the automaton as interpreting its own behaviors as generating obligations and partially fulfilling them or not doing so as the case might be.

Whether automata are constructible or exist in nature that may be accorded the status of rational agents rather than that of rational automata that conform to a logic of consistency by design or by natural law is an issue into which I shall not enter. I do take for granted that at least human beings are with relatively rare exceptions rational agents for much of their careers and that some social institutions sometimes are. Rovane (1994) has pointed out the interesting possibility that on some occasions so-called multiple personalities may be interpretable as several rational agents housed in a single human body. The point I mean to belabor here is that rational agents are understood as self critical agents who, among other things, are held up to the standard of rational health requiring consistency and logical closure of full belief and satisfaction of the BB principle.

The self critical dimension of rational agency demands that rational agents have still additional obligations. In order to identify the state of doxastic commitment, the agent is also committed to fully believing that *h* is not part of the doxastic commitment, if it is not.

Given any proposition *h,* X is committed either to full belief that *h,* full belief that $\sim h$ or to suspension of judgment between *h* and $\sim h$. No fourth commitment alternative is available. Thus, X is committed to fully believe that nine is the integer in the billionth place of the decimal expansion of π or to fully believe that it is not or to be in suspense on this issue. But only those who have obtained the printout of a program for generating this decimal expansion are capable of satisfying this commitment. Assuming that X is committed to full belief in theses of arithmetic, any behavior interpretable as suspense will violate X's commitments; for either it is an arithmetic thesis that nine is the integer or it is an arithmetic thesis that it is not. X might recognize this and refuse the posture of being in suspense without being able to make up his or her mind on whether nine is or is not the right integer. Nonetheless, all rational agents are committed to full belief, full disbelief or suspense on this issue at all times.

(B~B): If X at *t* is not committed to full belief that *h*, X at *t* is committed to fully believing that X does not fully believe at *t* that *h*.

As in the case of the BB principle, deliberating agents do fail to fulfill this B~B principle fully. An agent may fail to believe that he or she does not believe that *h* even though he or she does not have the commitment to believe that *h*. Lack of a commitment to fully believe that *h* can arise because the agent is committed to fully believing that ~*h* or because the agent is committed to being in suspense between *h* and ~*h* - that is, to judging both *h* and ~*h* to be serious possibilities. If in such a situation, the agent fully believes that he or she fully believes that *h*, the B~B principle is violated. Such violation may be due to lack of the requisite computational capacity. Or it may be attributable to some psychological disorder. These forms of self deception are just as much failures to satisfy the demands of the logic of consistency for full belief as is failure to satisfy the BB principle or the requirements of deductive consistency and closure. To understand X's full belief that X fully believes that *h* to be a failure to fulfill a commitment is to claim that the dispositions that would normally be interpreted to generate a commitment to believe that X believes that *h* cannot be so interpreted on this occasion but are better construed as a failed attempt to do so. As before, the remedy for this is to be sought through training, therapy and the use of prosthetic devices.

Imposing the B~B principle as a feature of the logic of consistency for full belief has seemed more controversial than imposing the BB principle. However, it must be imposed if we are to supplement the logic of consistency for full belief with a logic of consistency for probability judgment. A self reflective agent will be required in deliberation to identify his or her credal probability judgments including credal probability judgments assigning positive probability to hypotheses and to their negations. That is to say, the deliberating agent will be committed to full belief that his or her credal probability that *h* is positive if it is positive. But assigning positive probability that *h* commits the agent to judging *h* to be seriously possible - that is, to not fully believing that ~*h*. The agent who is committed to full belief that his or her credal probability that h is positive and, therefore, to not fully believing that ~*h* should also be committed to fully believing that he or she does not fully believe that ~*h*. Otherwise the agent cannot be using his or her state of full belief as a standard for serious possibility defining the space of possibilities over which credal probability judgments are made. The B~B principle seems to be required by appeal to the demands of self criticism extended to cover credal probability judgment.

The appeal to the requirement that rational agents are committed to the exercise of critical control over their own doxastic performance to determine how well they fulfill their doxastic commitments thus justifies supplementing the constraints on doxastic commitment generated by deductive logic with two additional constraints: the BB principle according to which an agent committed to fully believing that *h* is committed to fully believing that he or she fully believes that *h* and the B~B principle according to which an agent who is not committed to fully believing that *h* is committed to fully believing that he or she does not fully believe that *h*.

Commitment to conformity to the BB and B~B principles does not, however, generate a commitment to believe a certain set of extended logical truths additional to the truths of deductive logic in the way the principles of logical consistency and closure do. To come to grips with this point, we need to examine more closely what a logically true or valid proposition should be taken to be in developing a logic of consistency for full belief.

6

According to the account sketched initially, the standard of rational health for full belief spelled out by the requirements of logical closure and consistency require that all rational agents are committed to full beliefs that are closed under logical consequence and are consistent. This requirement demands that rational agents be committed to full belief in all logical truths no matter what other full beliefs they may have. This condition may be spelled out slightly more formally as follows:

The logical theses expressible in $\underline{L}$ are *positively valid* for agent X at *t* in the sense that no matter what potential corpus $\underline{K}$ in $\underline{L}$ is X's corpus at *t,* all of the theses of deductive logic expressible in $\underline{L}$ are in $\underline{K}$. Deductive closure ensures this to be so.

The logical theses expressible in $\underline{L}$ are *negatively valid for agent X at t* in the sense that the negation of a logical thesis is never a member of X's corpus at *t* no matter what that potential corpus might be. Consistency and positive validity requires this.

The logical theses expressible in $\underline{L}$ are *omnitemporally valid for agent X* in the sense that for given X they are positively (and, hence, negatively) valid for X at every time *t.*

The logical theses expressible in $\underline{L}$ are *universally valid* in the sense that they are omnitemporally valid for every agent X. The universally valid theses in $\underline{L}$ are those that are judged true by universal commitment.

Here "universal" means no matter who the agent is, no matter the time the beliefs are held and no matter the potential belief state the agent is in at the time - i.e., no matter what the agent's creed at the time is. So no one should resist acknowledging that they are true and, indeed, not only true but valid in the sense that they would be true no matter what might be the case.[6]

Extralogical theses expressible in L are neither universally valid, omnitemporally valid, positively valid nor negatively valid.

Consider now a language ML that contains L but also contains sentences of the form X believes at *t* that *h* (where *h* is a sentence in ML together with all the truth functional and, perhaps, quantifications over the variables X and *t*). A potential corpus in MK is a deductively consistent and closed set of sentences in ML that for each X and *t* satisfies the requirements of BB and B~B. X's corpus $K_{X,t}$ at *t* in L is to be understood as the intersection of X's corpus $MK_{X,t}$ at *t* in ML and L. As before, a sentence *g* in ML is in $MK_{X,t}$ if and only if X is committed at *t* to fully believe that *g*. Thus, if X is committed at *t* to fully believe that X fully believes at *t* that *h*, $MK_{X,t}$ contains "X fully believes at *t* that X fully believes at *t* that *h*". The distinction between negatively valid theses for X at *t*, positively valid theses for X at *t*, omnitemporally valid theses for X and universally valid theses carries over to sentences in ML.

If *h* (in L) is in $K_{X,t}$, the sentence "X believes at *t* that *h*" is in $MK_{X,t}$ in virtue of the BB principle. So is *h*. Hence, sentence BB(X,*t*,*h*) in ML that asserts that $h \supset$ X fully believes at *t* that *h* is in $MK_{X,t}$. If ~*h* is in K, the sentence BB(X,*t*,*h*) is still in $MK_{X,t}$. But BB(X,*t*,*h*) is not positively valid for X at *t*. When neither *h* nor ~*h* is in K, X at *t* is not committed to full belief that BB(X,*t*,*h*). Hence, BB(X,*t*,*h*) is not positively valid for X at *t*. BB(X,*t*,*h*) is negatively valid for X at *t*. Its negation is not in $MK_{X,t}$. That is to say, it would violate the standards of doxastic consistency for X at *t* to

6 I contend (Levi, 1996) that suppositional reasoning of the sort expressed in belief-contravening ("counterfactual") conditionals is to be represented by a transformation of a deductively closed and consistent corpus K in L to another such corpus K^{*r}_h. This transformation (Ramsey revision) is a variation of the more familiar AGM revision transformation (Alchourrón, Gärdenfors and Makinson, 1985). Supposing for the sake of the argument or fantasizing, no matter how fantastical it might be, cannot be coherent unless the transformation is from the current belief state meeting the requirements of the logic of consistency for full belief to another potential belief state meeting the same standards. Thus, any thesis positively valid for X at *t* should be in every belief state that is representable by a corpus obtainable via Ramsey revision of X's corpus $K_{X,t}$ by adding any consistent sentence *h* in L. A universally valid thesis has this property for every X and *t*. In this sense, a thesis judged true by universal commitment may be considered valid in the sense of being true in all possible worlds or under all interpretations or models whether they embrace Frege's objectivist view of logic or the more moderate view endorsed here.

fully believe that ~BB(X,*t*,*h*). Clearly BB(X,*t*,*h*) is neither omnitemporally valid for X nor universally valid. BB(X,*t*,*h*) does not qualify as a logical truth in any respect that the theses of deductive logic do except with respect to negative validity.

Consider, however, the sentence BB(X,*t*,X fully believes at *t* that *h*) that asserts the following in ML: X fully believes at *t* that $h \supset$ X fully believes at *t* that X fully believes at *t* that *h*. This is a sentence in $MK_{X,t}$ whether X is committed at *t* to full belief that *h*, full belief that ~*h* or to being in suspense. If X is committed at *t* to full belief that *h*, X is committed at *t* to full belief that X believes at *t* that *h*. That is required by the BB principle. "X fully believes at *t* that X fully believes at *t* that *h*" is in $MK_{X,t}$ if and only if X is committed at *t* to full belief that *h*. So "X fully believes at *t* that X fully believes at *t* that *h*" is in $MK_{X,t}$. Deductive closure then requires that BB(X,*t*,X fully believes at *t* that *h*) should also be in MK. In case X at *t* is committed to full belief that ~*h* or to suspense between *h* and ~*h*, X is committed not to believe that *h*. So *h* is not in MK. By deductive closure, BB(X,*t*,X fully believes at *t* that *h*) is in MK. It is positively valid for X at *t*. But it is not omnitemporally valid. Let MK' be X's corpus at *t*'. BB(X,*t*,X fully believes at *t* that *h*) need not be in MK' as representing X's part of X's commitment to full belief at *t*' concerning X's full beliefs at *t*. X at *t*' might be committed to doubt or disbelief as to whether X at *t* fully believes that X fully believes at *t* that *h* even though X at *t*' is committed to full belief that X at *t* fully believes that *h*. It is no part of the doxastic commitments of X at *t*' that he fully believe that X at *t* fulfills X's doxastic commitments at *t*. So BB(X,*t*,X fully believes at *t* that *h*) is not positively valid for X at *t*' and, hence, is not omnitemporally valid. *A fortiori*, BB(X,*t*,X fully believes at *t* that *h*) is not universally valid. BB(X,*t*,X fully believes at *t* that *h*) lacks the universal validity of a logical thesis even though it is positively valid for X at *t*.

To obtain a form of universality, we need to consider the BB principle itself. It characterizes constraints on the doxastic commitments of every rational agent at all times. But the BB principle is a *prescription* specifying the doxastic obligations of rational agents. All agents are committed to have full beliefs conforming to it. But commitment to full belief is here distinguished from the full beliefs that fulfill the commitment. The BB principle is a principle of normative doxastic logic but is not a universally valid proposition. Like the requirement that commitments to judgments of credal probability conform to the requirements of the calculus of probabilities, it is not a universally valid proposition even if it is a universally obligatory prescription. The requirements of deductive consistency and closure are also universally mandatory principles of the logic of consistency for full belief. They too are not universally valid propositions but unlike the BB principle they secure the universal validity of the theses of deductive logic and their truth by universal commitment.

Turn now to the B~B principle. If *h* is not in X's corpus K at *t*, it is not in $\underline{MK}_{X,t}$. However, X is committed at *t* to believe that he does not believe at *t* that *h*. So "X does not believe at *t* that *h*" is in $\underline{MK}_{X,t}$. So is the sentence "X does not fully believe at *t* that *h* ⊃ X fully believes at *t* that X does not fully believe at *t* that *h*". Call this B~B(X,*t*,X does not believe at *t* that *h*). This sentence is positively valid for X at *t*, but is neither omnitemporally nor universally valid. Yet, B~B is a principle of normative doxastic logic - that is, of the logic of consistency for full belief.

The upshot is that the BB and B~B principles do not generate commitments to full belief additional to the theses of deductive logic that, like logical theses, are universally valid in ML. Every rational agent is committed at every time to believe that logical theses are true. Every rational agent is committed at every time to avoid full beliefs inconsistent with such logical theses. But there are no additional propositions possessing truth by universal commitment even though the logic of consistency imposes doxastic obligations additional to the commitment to deductive closure and consistency. In this respect, therefore, the logic of consistency for full belief no more coincides with a logic of truth than the logic of consistency for credal probability judgment does *even if we overlook judgments of serious possibility and restrict ourselves to truth value bearing full beliefs.*

We can, to be sure, collect all the propositions that are positively valid for X at *t* and claim them to be logical truths for X at *t*. But logical truths for X at *t* may be falsehoods for X at *t*' or for Y at *t*. Talk of logical truths for an agent at a time grates. It is not surprising that this should be so. Logical truth is truth by universal commitment. If *h* is positively valid for X at *t*, it is true according to X's commitment at *t* no matter what potential corpus MK in ML satisfying the requirements of the logic of consistency is identical with $\underline{MK}_{X,t}$. But it does depend on both X and the time. Truth according to X's creed at *t* no matter what it is remains a far cry from truth by universal commitment.

Still theses positively valid for X at *t* reflect doxastic obligations X has at *t* no matter what X's creed at *t* might be. And these doxastic obligations are universalizable in the sense that Y at *t*' has doxastic commitments structurally similar to X's at *t* even if they are not identical. We might be prepared to accept some notion of truth according to X's creed at *t* no matter what it is as capturing logical theses if positive validity for X at *t* coincided with negative validity for X at *t*. If validity in the logic of truth is true no matter what might be the case, *h* is valid if and only if ~*h* could not be the case. Positive and negative validity coincide in the logic of truth.[7] They do not in the logic of consistency for full belief.

7

Consider Moore's paradox of saying and disbelieving. When X assents to the sentence "*h* but I, X, do not believe that *h*", X is a manifesting disposition to assent that might *prima facie* be interpreted as directly committing X to full belief that ~BB(X,*t*,*h*). But that violates the conditions for commitment. The behavior so interpreted is incoherent. BB(X,*t*,*h*) is negatively valid. X is prohibited by the logic of full belief from having a commitment to full belief at *t* that the negation is true. But X is not, thereby, committed to full belief that BB(X,*t*,*h*) by the logic of consistency for full belief. BB(X,*t*,*h*) is not positively valid. X is not committed at *t* to full belief that it is true. Moreover, it would be incoherent for him to be so committed no matter what *h* in ML might be involved.

The valid sentences of a logic of truth are supposed to be judged true by universal commitment. Part of the paradoxical seeming character of Moore's example derives from the fact that it is incoherent for X at *t* to be committed to full belief that ~BB(X,*t*,*h*) while there is no difficulty with Y's believing it at *t* or *t*'. But there is worse to come: I conjecture that much of the puzzlement concerning Moore's paradox derives from the failure of the positive and negative theses for X at *t* to coincide in the case of claims like BB(X,*t*,*h*) as they are expected to do if a logic of truth for full belief coincides with the logic of consistency for full belief.

Neither the relativity of validity to X at *t* nor the failure of positive and negative validity to coincide is paradoxical or even paradoxical seeming

7 As noted in footnote 2, I am assuming a classical nonintuitionistic logic of consistency for full belief. I also take for granted, therefore, that the logic of truth is a classic two valued logic as should be appropriate, in my judgment, to truth value as applied to belief. The case for proceeding in this way, like the argument of footnote 2, is based on the views advanced in Levi (1991, ch. 2). Shifting from the current belief state to one that is weaker incurs no risk of error. Every other shift does (p. 12). In particular, if an agent were in the weakest potential belief state, shifting to the join *h* and ¬*h* should incur a risk of error if negation is intuitionist; for such suspense is stronger than maximal ignorance. But there is no consistent belief state that falsifies the join of the states represented by these propositions. So the notion of risk of error makes no sense. The set of potential belief states ought to constitute at a minimum a boolean algebra reflecting the demands of a classical two valued logic. In spite of this rejection of intuitionism, positive validity and negative validity for an agent at a time do not agree in the logic of consistency for full belief.

when the logic of consistency is not conflated with the logic of truth. For those imbued with the traditions of Frege and Russell, Moore's paradox looks more troublesome simply because they have endorsed the conflation of the two. The obvious conclusion is: So much the worse for Frege's objectivism.

8

$\underline{MK}_{X,t}$ at *t* uniquely determines X's $\underline{K}_{X,t}$ at *t.* If *h* in $\underline{L}$ is in $\underline{MK}_{X,t}$ so that X is committed at *t* to fully believe that *h, h* is also in $\underline{K}_{X,t}$. This claim clearly does not generate any new commitments positively valid for X at *t.* It commits X to $h \supset h$, which is a universally valid thesis in deductive logic to which closure already commits X.

Does $\underline{K}_{X,t}$ determine $\underline{MK}_{X,t}$? I think it does.

Suppose that X is committed at *t* to fully believing that X fully believes at *t* that *h*. Given such commitment, it is generally conceded that X at *t* is committed to fully believing that *h.* We do not, however, need to assume this as an additional principle of the normative logic of consistency for full belief. We already have the resources to defend it. If X is not committed to believing that *h,* the OC principle (that is itself derivable from the requirements of logical consistency and closure, SSP and ~B) demands that either X is committed to believing that ~*h* or to suspending judgment. In either case, X is committed at t not to fully believe that *h.* The B~B principle then requires X to be committed to full belief that X does not believe that *h.* Since X is by hypothesis committed to full belief that X does fully believe that *h,* we have the conclusion that X's doxastic commitments violate the requirement of deductive consistency. $\underline{MK}_{X,t}$ contains both the claim that X does fully believe at *t* that *h* and its negation. Hence, the principles of consistency and closure together with B~B require that X's doxastic commitments at *t* satisfy the following BT condition:

If X at *t* is committed to full belief that X at *t* believes that *h,* then X is committed at *t* to full belief that *h.*

The BT principle requires that if "X believes at *t* that *h*" is in $\underline{MK}_{X,t}$ at *t,* *h* is in $\underline{K}_{X,t}$ and, hence, in $\underline{MK}_{X,t}$. Consequently, BT(X,*t*,*h*) (= "X believes at *t* that $h \supset h$") is in $\underline{MK}_{X,t}$. When "X does not believe at *t* that *h*" is in $\underline{MK}_{X,t}$, BT(X,*t*,*h*) will still be present.

Hence, BT(X,*t*,*h*) is positively valid for X at *t.* It is not, however, omnitemporally valid or universally valid for reasons paralleling those offered for the positively valid propositions supported by BB and B~B.

That is to say, if we consider the corpus $\underline{MK}^*$ representing X's corpus at t' or Y's corpus at t, the BT(X,t,h) need not be present in $\underline{MK}^*$.

Thus, the BT principle does not add any further universally valid logical theses. It does yield new positively valid theses to X's corpus at t. And it does something more. It allows us to say that X's corpus $\underline{K}_{X,t}$ at t expressible in $\underline{L}$ uniquely determines that portion of his corpus in $\underline{MK}_{X,t}$ that represents X's doxastic commitments as to what he believes - insofar as this does not pertain to other agents' doxastic commitments. When it comes to the commitments of other agents (say, agent Y's commitments at t' to full belief), the set of constraints introduced allows unique determination from X's commitments at t to what the contents of Y's corpus $\underline{MK}_{Y,t'}$ in $\underline{ML}$ that are not expressed with an initial "Y believes at t that" operator. This includes Y's commitments to full belief at t' as to what X's full beliefs at t or some other time are as well as Y's commitments to full belief representable by sentences in $\underline{L}$.

Last but not least, the requirements of consistency and closure on potential corpora in $\underline{ML}$ together with the BB, B~B, SSP and ~B principles guarantee that every consistent potential corpus in $\underline{ML}$ contains as positively valid formulas for X at t theses of an S5 modal logic using "X fully believes at t that" as the necessity operator. In this sense, the logic of consistency for full belief yields an S5 doxastic logic for the full beliefs of X at t. It yields a different S5 doxastic logic for the full beliefs of X at t' and for Y at both t and t'. In all cases, the structure of the doxastic logic is the same, but different theses qualify as theses of the S5 doxastic logic. Except for the theses of deductive logic, there are no universally valid theses. Except for the theses of deductive logic, there are no truths by universal commitment of S5 doxastic logic. To have a coherent or consistent system of beliefs, X's beliefs should conform to S5 requirements. But consistency of full belief is a far cry from truth of full belief.

According to the account offered earlier, h is in $\underline{MK}_{X,t}$ if and only if X is committed to full belief at t that h. h is not in $\underline{MK}_{X,t}$ if and only if X is committed at t not to believe that h. The presence of h in $\underline{MK}_{X,t}$ does not imply that h is true (although it does imply that X is committed to full belief at t that h is true). The absence of h in $\underline{MK}_{X,t}$ does not imply anything about the truth or falsity of h either. It implies only that X at t is committed not to believe that h. The only issue of fact is whether X at t has dispositions requisite to generate commitments to full belief at t in the truth of the items listed in $\underline{MK}_{X,t}$ and has commitment not to believe items that do not belong to that set. The logic of consistency characterizes conditions every such corpus for every X at every time should satisfy. None of these conditions stipulate that any theses positively valid for an agent X at *t* in $\underline{\text{ML}}$ are true except, perhaps, for the truths by universal commitment or theses of deductive logic. In particular, BT(X,*t*,*h*) is not dictated by the logic of consistency for full belief to be true. X and X alone at *t* is rationally obligated to fully believe that BT(X,*t*,*h*). No one, not even X, is committed to do so at any other time. Still philosophers, computer scientists, game theorists and other would-be users of doxastic logic deny that the logic of full belief is S5.

Now it is often an affront to the sensibilities of others to declare all of one's own beliefs to be true. Political correctness in philosophical discourse leads to confusion just as directly as it does in other contexts. I do not think that proper speech is the only concern here – or, at least, I would like to think that something philosophically more important is at stake. My conjecture is that resistance to claiming that the logic of consistency for full belief for an agent at a time yields an S5 structure for beliefs positively valid for the agent at that time is a concern on the part of those who think we should have the logic of consistency for full belief coincide with the logic of truth for full belief. I believe I have said enough already to establish the futility of the effort; however, I have done so without mentioning Kripke semantics for doxastic or epistemic logic where, so it might be hoped, salvation of the Frege-Russell vision of logic might be sought when the logic of belief is considered a logic.

9

For any sentence *h* in $\underline{\text{ML}}$, *h* is in $\underline{\text{MK}}_{\text{X},t}$ if and only if X at *t* fully believes that *h*. The sentence "X at *t* fully believes that *h*" (which is also a sentence in $\underline{\text{ML}}$) is in $T(\underline{\text{MK}}_{\text{X},t})$ if and only if X at *t* fully believes that *h*. Hence, *h* is in $\underline{\text{MK}}_{\text{X},t}$ if and only if "X fully believes that *h*" is in $T(\underline{\text{MK}}_{\text{X},t})$. $T(\underline{\text{MK}}_{\text{X},t})$ is precisely the set of sentences $\underline{\text{MK}}_{\text{X},t}/\underline{\text{K}}_{\text{X},t}$ in $\underline{\text{ML}}/\underline{\text{L}}$; but membership in this set is interpreted in two distinct ways. When the set is named by "$T(\underline{\text{MK}}_{\text{X},t})$," every sentence in the set is asserted to be true. The elements of $T(\underline{\text{MK}}_{\text{X},t})$ specify what X does or does not fully believe at *t*. Instead of representing X's belief state at *t* in this fashion, $\underline{\text{MK}}_{\text{X},t}$ lists each sentence *h* in $\underline{\text{ML}}$ representing one of X's full beliefs at *t*. In the same spirit, $\underline{\text{MK}}_{\text{X},t}/\underline{\text{K}}_{\text{X},t}$ lists those sentences in $\underline{\text{ML}}/\underline{\text{L}}$ that represent X's full beliefs about what X does and does not fully believe, but this list does not include those sentences that represent those full beliefs that are not about X's beliefs or disbeliefs. When X's belief state is represented by a set of sentences constituting a list of sentences in $\underline{\text{ML}}$, the representation is $\underline{\text{MK}}_{\text{X},t}$. When X's belief state is represented by a set of sentences asserting what X's beliefs are, the set is $T(\underline{\text{MK}}_{\text{X},t}) \subset \underline{\text{MK}}_{\text{X},t}$.

Consider, then, the deductive closure $S(T(\underline{\mathrm{MK}}_{\mathrm{X},t}),\underline{\mathrm{K}}_{\mathrm{X},t})$ of the union of $T(\underline{\mathrm{MK}}_{\mathrm{X},t})$ and $\underline{\mathrm{K}}_{\mathrm{X},t}$ where $T(\underline{\mathrm{MK}}_{\mathrm{X},t})$ represents X's belief state at t as given above and h in $\underline{\mathrm{L}}$ is in $\underline{\mathrm{K}}_{\mathrm{X},t}$ if and only if h is true. This set is identical with $\underline{\mathrm{MK}}_{\mathrm{X},t}$, but the set is now being used with another intended interpretation. $T(\underline{\mathrm{MK}}_{\mathrm{X},t})$ is used to describe X's state of full belief at t. If h is in $\underline{\mathrm{L}}$, "X at t believes that h" is in $T(\underline{\mathrm{MK}}_{\mathrm{X},t})$ and h (or "it is true that h") is in $\underline{\mathrm{K}}_{\mathrm{X},t}$. So $S(T(\underline{\mathrm{MK}}_{\mathrm{X},t}),\underline{\mathrm{K}}_{\mathrm{X},t})$ describes X's state of belief at t and asserts that it is free of error.

Like $\underline{\mathrm{MK}}_{\mathrm{X},t}$, $S(T(\underline{\mathrm{MK}}_{\mathrm{X},t}),\underline{\mathrm{K}}_{\mathrm{X},t})$ must have an S5 structure. The S5 structure of $\underline{\mathrm{MK}}_{\mathrm{X},t}$ is required by the logic of consistency for full belief. X at t is committed to full belief that BT(X,t,h). X is obliged at t to judge BT(X,t,h) true, but no one else is. When we deploy $S(T(\underline{\mathrm{MK}}_{\mathrm{X},t}),\underline{\mathrm{K}}_{\mathrm{X},t})$, not only do we claim that X at t is committed to judging BT(X,t,h) to be true but we now judge BT(X,t,h) to be true. The logic of consistency requires that X make the judgment. It does not oblige us to do so.

Take any potential corpus in $\underline{\mathrm{MK}}_{\mathrm{X},t}$ in $\underline{\mathrm{ML}}$ and introduce $S(T(\underline{\mathrm{MK}}_{\mathrm{X},t}),$ $\underline{\mathrm{K}}_{\mathrm{X},t})$. For any such $\underline{\mathrm{MK}}_{\mathrm{X},t}$, consider all maximally consistent extensions in $\underline{\mathrm{L}}$ of $\underline{\mathrm{K}}_{\mathrm{X},t}$. $S(T(\underline{\mathrm{MK}}_{\mathrm{X},t},w)$ for any such extension w represents a "possible world" in which X's state of full belief is specified by $\underline{\mathrm{MK}}_{\mathrm{X},t}$ and all of X's full beliefs are true. All the sentences that are instances of theses of S5 modal logic are true under the assumptions we are making no matter what the contents of the original $\underline{\mathrm{MK}}_{\mathrm{X},t}$ happen to be. That is to say, the theses of an S5 modal logic in the full $\underline{\mathrm{ML}}$ can be shown to be universally valid or true in every $S(T(\underline{\mathrm{MK}}_{\mathrm{X},t}, w)$ or possible world constructed in this manner. Each possible world specifies a state of doxastic commitment for X at t. The alternatives to each such possible world z are precisely those possible worlds in which that state of doxastic commitment by X at t that holds at z obtains. We have a species of semantic models structured along lines pioneered by Kanger (1957), Hintikka (1969, section III) and Kripke (1963) in variant ways. The theses of S5 are now valid (i.e., true in all possible worlds) as a logic of truth might require when alternativeness is an equivalence relation. Equating the logic of truth with the logic of consistency would require that every rational agent Y at every t' is committed to believing all of these theses true.

If the S5 doxastic logic were interpretable as both a logic of consistent or coherent full belief and a logic of truth, some vision such as this would be required. The vision is, of course, a sheer fantasy. It requires every rational agent to fully believe that what anyone at any time fully believes is true.

The sensible response to this absurdity is to abandon the identification of the logic of truth with the logic of consistency for full belief. For the reasons offered previously, an S5 logic of consistency should be retained to specify constraints on belief states represented by potential corpora $\underline{\mathrm{MK}}_{X,t}$ in $\underline{\mathrm{ML}}$. In order to obtain a logic of truth, let maximally consistent sets *w* of sentences in $\underline{\mathrm{L}}$ be allowed to be arguments in $S(T(\underline{\mathrm{MK}}_{X,t}), w)$. The maximally consistent sets in $\underline{\mathrm{ML}}$ [represented by the logical consequences of the union of $T(\underline{\mathrm{MK}}_{X,t})$ and *w* for every $\underline{\mathrm{MK}}_{X,t}$ satisfying S5 in $\underline{\mathrm{ML}}$ and every maximally consistent *w* in $\underline{\mathrm{L}}$)] would then yield a set of sentences true in all possible worlds exhibiting the structure KD45 or S5-manqué that is not committed to the validity of sentences BT(X,*t*,*h*) unless *h* is a sentence possessing a prefix of the form "X at *t* believes that" or "X at *t* does not believe that".

Of course, this "logic of truth" is so called only with some charity; because the only "possible worlds" contemplated are ones where the agent X actually fulfills the requirements of the S5 logic of consistency for full beliefs - a condition which it is incredible to suppose any agent X satisfies. At best, we have a model for ideally rational agents and not a logic of truth. Even if such "worlds" are logically possible, there are worlds where the requirements specified fail. This observation need not, however, deter us too much if we claim that the logic of truth is counterfactual focusing on what might be true in any situation where all the agents concerned were ideally rational.[8]

The main point is that even with this charitable construal, the logic of consistency specifying what a rational agent ought to fully believe to be ideally consistent remains S5 and, hence, different from the S5-manqué logic of truth. Not only in the case of probability judgment but in the logic of full belief, there is no coincidence between the logic of truth and the logic of consistency.

Weakening the logic of consistency to S5-manqué while keeping the S5-manqué logic of truth intact is not workable. BT(X,*t*,*h*) will no longer be positively valid for X at *t*. Hence, BT(X,*t*,X believes at t that *h*) will not be in $T(\underline{\mathrm{MK}}_{X,t})$ in some possible cases even though it remains in $\underline{\mathrm{MK}}_{X,t}$ and is positively valid for X at *t*. Hence, some sentences positively valid for X at *t* will not be valid in the sense of being true in all possible

8 Even if one has qualms about regarding this structure as a logic of truth, it is a logic in the structural sense of Koslow (1992) where we have a classical implication structure and its dual with negation, disjunction and conjunction and where "X fully believes at *t* that" and "X does not fully believe that ~" are modal operators with respect to the implication structure and its dual that are Y modals (section 35.4), K_4 modals (section 35.3) and S5 modals (section 35.9). The assumptions we have specified to hold ensure that the conditions for such an implication structure are satisfied so that we have a logic in Koslow's structural sense. What one may well doubt is that the implication structure so generated is suited for characterizing a logic of truth.

worlds. We can no longer assume following the approach adopted above that BT(X,*t*, X believes at *t* that *h*) will be in $S(\underline{MK}_{X,t},w)$ for every possible belief state for X at *t* and every *w*. Both the logic of consistency and the logic of truth for full belief will be weakened in a way that prevents coincidence of the two.

Moreover, weakening the S5 logic of consistency to S5-manqué or something else entails giving up either SSP, ~B, or B~B in a way that abandons the idea that the state of full belief defines the space of serious possibility over which credal probability judgments are made. Or at least this cannot be done without abandoning the use of standard requirements that for probability judgment to be coherent the requirements of a finitely additive probability measure must be satisfied by numerically precise probability judgments.

For X at *t* to fully believe that what X at *t* fully believes is true smacks of doxastic immodesty. If so, it is immodesty mandated by doxastic coherence. I have been arguing that one cannot give up this coherence lightly. In any case, the demand for such coherence does *not* require that X be committed to full belief that X has fulfilled the demands imposed by X's commitments to full belief. Such arrogance would be monstrous. Moreover, X's rationally mandatory doxastic immodesty need not be accompanied by Y's credulity. Y is not obligated as a rational agent to full belief that whatever X fully believes at some specific time is true.

10

If, as I have been arguing, the logic of full belief characterizing an agent's commitments to full belief at *t* is S5, and insofar as we may ignore X's commitments to full belief regarding other agents, $\underline{MK}_{X,t}$ is uniquely determined by $\underline{K}_{X,t}$. Consequently, changes in X's doxastic commitments can be represented as changes in X's corpus $\underline{K}$ in $\underline{L}$. This point is important to keep in mind when exploring the ways in which a doxastic commitment may be modified. In particular, if neither *h* nor ~*h* in $\underline{L}$ is in $\underline{K}_{X,t}$, $\underline{K}_{X,t}$ can be changed by adding *h* and forming the deductive closure. Expansion of the initial corpus by adding *h* (perhaps through observation, by receiving information from others or by ampliative inference) cannot be represented formally as an expansion if *h* is added to $\underline{MK}_{X,t}$ and taking the deductive closure. The transformed corpus in $\underline{ML}$ will not represent the doxastic commitments generated by the corresponding corpus in $\underline{L}$. Indeed, there is no way that $\underline{MK}_{X,t}$ can be expanded with doxastic consistency. In virtue of the S5 structure of the logic of consistency for full belief, we may happily focus on expansion of corpora in $\underline{L}$.

Similar observations apply *mutatis mutandis* to suppositional reasoning, the understanding of conditional judgments of serious possibility and the logic of conditionals that regulates the coherence of such judgments (Levi, 1996).

11

Shifting gears somewhat, what may we say about epistemic logic or the logic of knowledge? Much depends upon how one understands knowledge. According to the pragmatist view I favor, what X knows is, according to what X fully believes, what X fully believes; for what X fully believes, X is committed to judging to be true. And it is common ground among the classical pragmatists that, from X's point of view, what X fully believes (does not seriously doubt) stands in no need of justification. Justification is required for changing one's full beliefs and not for the full beliefs one has. Since the logic of consistency of full belief for agent X at *t* requires X to equate what X is committed at *t* to know with what X is committed at *t* to fully believe, there can be no logic of consistency for knowledge distinct from the logic of consistency for full belief. The positively valid theses for X at *t* of the logic of consistency for knowledge are the same S5 theses as apply to the logic of consistency for full belief.

Even if Y at *t*' fully believes that X at *t* completely satisfies the requirements for doxastic and, hence, epistemic coherence, Y's views regarding X's full belief need not equate them with X's knowledge. Y will be committed to the supposition that X's knowledge satisfies a KK principle corresponding to BB and a KT principle but not with a K~K condition.[9] Suppose that Y fully and correctly believes that X's full beliefs at *t* satisfy the S5 requirements for doxastic consistency. If Y fully believes that X at *t* fully believes that *h* and also that *h* is false, Y should fully believe that X fully believes at *t* that X fully believes at *t* that *h* and, indeed, fully believes at *t* that X knows at *t* that *h*. But Y believes that X does not know at *t* that *h* because Y judges *h* to be false. Y also believes that X does not know at *t* that X does not know that *h*. So the K~K condition fails. That is to say, K~K fails as a model of Y's beliefs concerning X's knowledge at *t*. But insofar as epistemic logic, like doxastic logic, is a logic of consistent evaluation by an agent X concerning what X knows at *t* or evaluation by agent Y of X's consistency in making such evaluations, the logic of full belief coincides with the logic of knowledge and both are S5.

Of course, the claim that X knows that *h* at *t* may mean even for a

9 See Hintikka (1962) for a development of an epistemic logic without K~K and Lamarre and Shoham (1994) for a counterinstance to K~K.

pragmatist that X not only truly believes that *h* at *t* but does so authoritatively in the sense that in some way or other he can justify to others the truth of what he fully believes. But whether X can do so or not even by X's own lights depends upon what by X's lights the others already fully believe. We should be skeptical of the availability of a logic of consistency for knowledge as distinct from the logic of consistency for full belief and even more skeptical of a useful epistemic logic of truth.

12

Suppose that X and Y at *t* share a common corpus $\underline{K}$ in $\underline{L}$ and both fully believe that this is so. Finally, they both fully believe that they are both doxastically coherent in the S5 sense I have been advocating. For each *h* in $\underline{K}$, X is committed by S5 to "X believes at *t* that *h*" in $\underline{MK}_{X,t}$ as well by the suppositions made above to "Y believes at *t* that *h*". Y is committed by parallel reasoning to the membership of these sentences in $\underline{MK}_{Y,t}$. By S5, X is also committed to "X believes at *t* that *h*" and to X believes at *t* that Y believes at *t* that *h*". By X's full belief at *t* that Y is doxastically consistent, X is also committed to full belief represented by "Y believes at *t* that X believes at *t* that *h*" and "Y believes at *t* that Y believes at *t* that *h*". And by parallel reasoning, Y is committed to these at *t* as well. It is easy now to see that iteration of this process can be carried on indefinitely. Both X and Y have a commitment to common belief that *h* at *t* and, indeed, to all elements in $\underline{K}$. Moreover, their common belief corpus has an S5 structure where what is positively valid for X is positively valid for Y.

In this case, we may, indeed, say that BB(X,*t*,X believes at *t* that *h*), B~B(X,*t*,X does not believe at *t* that *h*) and BT(X,*t*,*h*) are positively valid at *t* for X and Y. From their common perspective, their common full belief is common knowledge.

With this understanding, if everyone at all times regarded everyone else at all times to be doxastically coherent, and believed that everyone else always regarded them to be doxastically coherent, they all would be committed to common knowledge of the universally valid theses of deductive logic.

Of course, if Z outside the consensus fully believed that *h* is false, Z would deny that X and Y have common knowledge that *h* even though Z believed that they have common belief and that the common belief is doxastically consistent in the S5 sense even though it is false. It is an old commonplace that consistency does not imply truth. The main burden of this paper has been to suggest that those who do not think that the logic of consistency for full belief is S5 are so wedded to the coincidence of the logic of consistency with the logic of truth that they have forgotten this commonplace.

REFERENCES

Alchourrón, C. Gärdenfors, P. and Makinson, D. (1985). "On the Logic of Theory Change, Partial Meet Functions for Contraction and Revision," *Journal of Symbolic Logic,* 50: 510–30.

Frege, G. (1967). *The Basic Laws of Arithmetic,* translated and edited by M. Furth. Berkeley: University of California Press.

Hintikka, J. (1962). *Knowledge and Belief.* Ithica, N.Y.: Cornell University Press.

(1969). *Models for Modalities.* Dordrecht: Reidel.

Kanger, S. (1957). *Provability in Logic.* Stockholm: Almqvist and Wiksell.

Koslow, A. (1992). *A Structural Theory of Logic.* Cambridge: Cambridge University Press.

Kripke, S. (1963). "Semantical Considerations on Modal Logics," *Acta Philosophica Fennica, Modal and Many Valued Logics,* pp. 83–94.

Lamarre, P. and Shoham, P. (1994). "Knowledge, Certainty, Belief and Conditionalization" (unpublished).

Levi, I. (1991). *The Fixation of Belief and Its Undoing.* Cambridge: Cambridge University Press.

(1996). *For the Sake of the Argument.* Cambridge: Cambridge University Press.

Ramsey, F. (1990). *Philosophical Papers.* Edited by D. H. Mellor. Cambridge: Cambridge University Press.

Rovane, C. (1994). "The Personal Stance," *Philosophical Topics,* 22: 351–396.

4

Consequentialism and sequential choice

1. EXTENSIVE AND NORMAL FORM

P. J. Hammond opens his important 1988 paper by writing:

> An almost unquestioned hypothesis of modern normative decision theory is that acts are valued by their consequences. Indeed, Savage (1954) defines an act as a function mapping uncertain states of the world into a domain of conceivable consequences, thus identifying an act with the state-contingent consequence function which it generates. (Hammond, 1988, p. 25)

Hammond thinks that much of the controversy concerning acceptable principles of rational behavior (such as, for example, the status of the requirement that preferences should be weakly ordered[1] and should satisfy the independence postulate)[2] could be resolved if we appreciated

Thanks are due to Michael Bacharach and Susan Hurley for extensive and helpful comments. I wish to thank the Fellows of All Souls College for providing me with the leisure to work on the materials from which this paper emerged.

Published originally in a modified form in M. Bacharach and S. Hurley, eds., *Foundations of Decision Theory* (Oxford: Blackwell, 1991), pp. 92–122. Reprinted by permission of Blackwell Publishers.

1 Preference is to be understood as a propositional attitude comparing propositions as better or worse according to the agent's value commitments. Such comparisons may take into account political, economic, prudential, moral, cognitive or personal concerns. The requirement that the propositions in the domain under comparison be weakly ordered implies that for every pair of propositions in the domain either one of them is ranked above (is strictly preferred) the other or they are ranked together (are equipreferred or equivalued). The requirement of weak ordering rules out cases of noncomparability.

2 By speaking of "the independence postulate" I mean to allude to a family of related assumptions such as Savage's (1954) "sure thing principle" or the monotonicity and substitutability axioms of Luce and Raiffa (1958, pp. 27–28) that impose a constraint on preferences over propositions describing options available to the decision maker. The technical details will not be required in this discussion but roughly speaking it is a variant on principles prohibiting the choice of an option when there is another available that dominates it. For example, a dominates b if a is better than b in all possible situations according to at least one way of identifying a set of exclusive and exhaustive possible situations. Alternatively, we may be given a stock of prizes which the agent weakly orders with respect to his preferences and say that a stochastically dominates b if and only if the probability of obtaining a prize no better than x is less if a is chosen than if b is chosen for some x and is no greater for every x. Among the

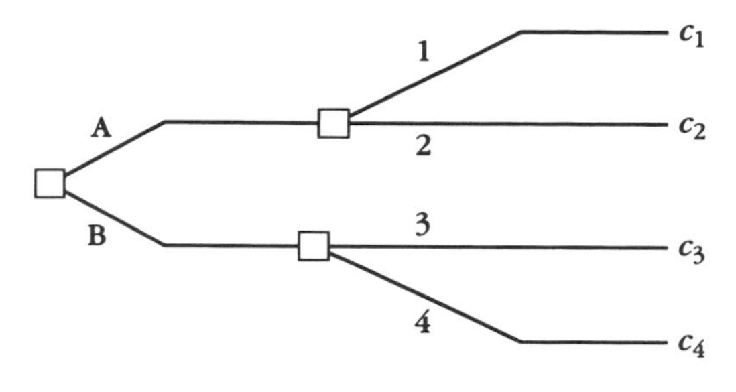

Figure 4.1

the implications of the "almost unquestioned" consequentialism lying at the core of modern normative decision theory for sequential decision making.

When an agent faces a choice among a set of available options, some or all of them may present him with fresh opportunities for choice among options which, in turn, may yield new opportunities for choice and so on. The decision problem is representable in a "tree form" or "extensive form" illustrated by Figure 4.1.

At stage 1, the agent has a pair of options A and B. Option A has as a sure consequence that the agent confronts a choice between 1 and 2. Option B has as a sure consequence that the agent confronts a choice between 3 and 4. In more sophisticated examples, the sequences of stages at which opportunities for choice arise can ramify. In addition, considerations of uncertainty and risk can be added. The c_i's are the consequences.

When the decision problem characterized by Figure 4.1 is recast in normal or strategic form, the agent is represented as being confronted at the initial stage with several plans or strategies specifying for each choice node which option at that choice node the agent takes and, in this sense, identifying a path in the "decision tree" of Figure 4.1. The normal form

several technical issues which complicate the formulation of a simple, neutral and accurate version of this principle is the circumstance that given the same pair of options a and b, a may (stochastically) dominate b relative to one way of partitioning into possible situations (possible prizes) but not relative to another. Advocates of the independence postulate insist that if there is at least one way of partitioning relative to which a (stochastically) dominates b, a should be preferred to b. Critics of the independence postulate, such as M. Allais (1953) and M. Machina (1982), in effect, relativize dominance requirements to privileged partitions into possible situations (prizes). For a survey of some of the formal variants of the independence postulate, see MacCrimmon and Larsson (1979).

representation glosses over the sequential aspects of the agent's predicament and views him as if he were making a single choice at the initial stage. Thus, the normal form representation corresponding to Figure 4.1 is given in Figure 4.2.

The normal form version of the predicament abstracts away from the sequential features of the decision problem revealed in Figure 4.1. The options such as A1 in Figure 4.2 could be available for the agent to choose at the initial node without opportunity to renege at some later stage. Alternatively, however, A1 could represent a strategy or a program for sequential choice endorsed at the initial node but open for reconsideration with the option of reneging at a later stage. These situations are quite different. Figure 4.1 disambiguates in favor of the second reading.

Hammond contends that a consequentialist who ignores the sequential aspects of the agent's decision problem revealed in the extensive form representation and pictures the predicament as in Figure 4.2 will understand the problem as one of choosing among the available consequences. The options admissible for choice will be precisely those determining admissible consequences. This is not big news.

Hammond thinks that in those situations where a Figure 4.2 representation is an alternative to a Figure 4.1 representation abstracting away from the details of sequential choice, consequentialism yields more substantial results. It tells us that the decision maker is rationally entitled to pursue a course of action (sequence of choices) if and only if it corresponds to an option in the normal form representation yielding an admissible choice of a consequence. The criteria for choice in the extensive form sequential decision problem are thereby derived from consequentialist criteria for a single choice normal form decision problem. Hammond argues that given this reduction of extensive to normal form, rational

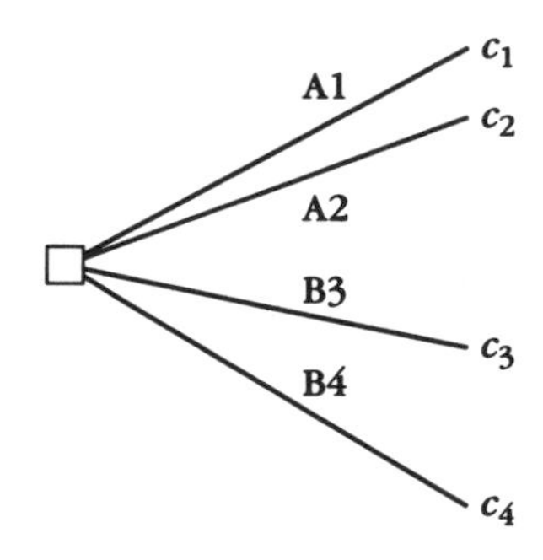

Figure 4.2

preference should induce a weak ordering among options and should satisfy independence. Consequentialism is thereby seen as supporting two pillars on which advocates of the injunction to maximize expected utility have often rested their position.

My aim is to call into question the contention that consequentialism has the implications Hammond claims it has. I am not suggesting any reason to doubt the cogency of Hammond's proofs. Furthermore, I am inclined to agree with Hammond that not only is consequentialism a widely endorsed assumption of modern normative decision theory but so is the reduction of extensive to normal form. I object to Hammond's assumption that consequentialism in the sense in which it is endorsed by Savage et al. presupposes the reduction of extensive to normal form and to the implication which Hammond correctly derives from this assumption that consequentialism entails ordinality.

The matter is of some importance. Although both consequentialism and the thesis of the reducibility of extensive to normal form are widely endorsed among contemporary decision theorists, the consequentialism invoked by "modern normative decision theory" is less controversial than the reducibility of extensive to normal form.[3] Those who reject the latter need not reject the former. Those who accept the former need not accept the latter. I, for one, have gone on record as rejecting the ordering conditions on preferences favored by Savage et al. (Levi, 1974, 1980, 1986a). At the same time, I am in favor of adhering to the independence postulate (Levi, 1986a, 1986b) given the satisfaction of certain structural conditions and am convinced by Seidenfeld's argument that consequentialism does sustain this conclusion (Seidenfeld, 1988). If Hammond is right, this position is untenable. Ordering and independence are indivisible. I think Hammond is wrong.

2. ACT, STATE, CONSEQUENCE

According to the procedure Savage adopted as canonical for representing decision problems, three notions are deployed in the representation: the notion of an act, a state and a consequence. Many philosophers have followed the lead of R. C. Jeffrey in complaining about a wrong headed

3 Although the consequentialism endorsed by "modern normative decision theory" is less controversial than the reduction of extensive to normal form, consequentialism has been a focal point of objections to utilitarian and utilitarian-like ethical theories. In this essay, I shall not be addressing the issues raised in this debate. However, by identifying the sense in which consequentialism is presupposed by "modern normative decision theory", we may obtain as a by-product a more just appreciation of the extent to which contemporary decision theory is or is not caught up in this controversy.

ontology which insists on trinitarianism where monotheism should do. Jeffrey (1965) suggests that acts, states and consequences are all events or propositions.

I do not want to quarrel with Jeffrey's suggestion. To me, something rather like it should turn out right. But I do not see why it should be supposed, as Jeffrey does intimate, that Savage would disagree or, for that matter, that Ramsey would disagree. Perhaps, Ramsey and Savage may be convicted of what now seems like loose talk; but it is only loose talk easily repaired without damage to the substance of their views. Instead of speaking of acts, states and consequences, Savage could have spoken of act-descriptions, state-descriptions and consequence-descriptions.

There are, to be sure, important differences in a Savage framework in the attitudes the decision maker has towards act-descriptions, state-descriptions and consequence descriptions. State-descriptions are objects of personal probability judgements (I prefer calling them "credal" probability judgements), consequence-descriptions are objects of utility judgement and act-descriptions of expected utility judgement. There is nothing in the Savage system to prevent assigning utilities to state-descriptions or probabilities to consequence-descriptions. Indeed, it seems clear that Savage intended the state-descriptions to be evaluated with respect to utility in a certain way although, as is well known, his axioms do not capture his intent. However, the utility assigned a state must be conditional on the act chosen. That is to say, given that act a_i is chosen and state s_j is true, the utility assigned to s_j conditional on a_i is equal to the unconditional utility of the consequence c_{ij}. Unless the consequences of all available options in a given state bear equal utility, the only way to derive an unconditional utility for the state is to compute the expectation of the conditional utilities of the state (unconditional utilities of the consequences) utilizing unconditional probabilities for acts. Similarly credal probabilities are assignable to consequences; but these are conditional on the option chosen. Unless consequences are identical for all options in a given state, the only way to compute unconditional probabilities for consequences is with the aid of unconditional probabilities for those options which yield them in some state or other.

These restrictions derive from the fact that in Savage's formalism, unconditional credal probabilities are not assigned to acts – i.e., the agent's options. According to the Savage approach, one begins with a preference or value ranking of acts or options which satisfies the axioms he proposes. From this ranking, one may derive a unique unconditional credal probability distribution over the states and an unconditional utility function unique up to a positive affine transformation over the consequences which is state independent. Utilizing this information, it is possible to derive a probability distribution over consequences conditional on acts and a utility function for states conditional on acts (the utility of state s_j conditional on act a_i is the utility of consequence c_{ij}). However, the Savage theory fails to determine an unconditional credal probability distribution over acts and, as a consequence, an unconditional utility function for states and an unconditional probability distribution over consequences.

To be sure, the Savage axioms, even when construed prescriptively as norms of rationality, are designed so that one might elicit from information about an agent's preferences among acts information about his probabilities and utilities. The fact that information about preferences among acts unsupplemented with other information fails to yield information about unconditional probabilities of acts need not imply that the agent fails to make such probability judgements. Nonetheless, even if it did have this implication, the further thesis that acts, states and consequences are distinct types of entities would not follow. By replacing talk of acts, states and consequences by talk of act-descriptions (act-propositions), state-descriptions (state propositions) and consequence-descriptions (state propositions), the differences between acts, states and consequences are seen to be differences between propositions or descriptions deriving from differences in the propositional attitudes which the decision maker may have towards act-descriptions, state-descriptions and consequence-descriptions.[4] The same sentence or prop-

4 A whiff of causality may seem to be introduced into our discussion owing to the recognition of the distinction between act-descriptions and consequence- and state-descriptions. Act-descriptions are sentences or propositions whose truth values are subject to the agent's control. By his choice, their truth values are settled. Or, more strictly speaking, propositions which are act-descriptions according to agent X are propositions whose truth values are subject to X's control according to X's corpus of certainties $\underline{K}$.

However, no matter what a deeper analysis of the concept of control might reveal, as far as decision theory is concerned, the epistemic condition X satisfies when we say the truth of act-descriptions is under X's control is this: The "expansion" $\underline{K}$ + "X chooses a_i" of X's corpus $\underline{K}$ by adding the information that X chooses a_i contains a_i – i.e., entails the truth of a_i while $\underline{K}$ does not. We may also say that the *truth value* of a proposition h that is not an act description is under X's control according to $\underline{K}$ if and only if there is some a_i such that $\underline{K} + a_i$ entails h and some a_k such that $K + a_k$ entails $\sim h$. The *truth* of h is under X's control according to $\underline{K}$ if and only if there is an a_i such that $\underline{K} + a_i$ entails h and another a^* such that $\underline{K} + a^*$ entails neither h nor $\sim h$.

This characterization of control of truth values proceeds along entirely epistemic lines once we are given a list of option-descriptions. Causality does not enter into the picture. In this sense, agents are in control of which of their options they choose and which of the "sure" consequences of their options will be realized. They are not in control of unsure consequences or of states.

osition may qualify as a state-description relative to one network of propositional attitudes and as a consequence-description or, indeed, even an act-description relative to another. We may, therefore, endorse Jeffrey's insistence on regarding acts, states and consequences propositionally without accepting his contention that the decision maker should assign "desirabilities" (utilities or expected utilities) and probabilities over propositions of these three kinds.

As W. Spohn (1977, 1978) has rightly recognized, we should resist Jeffrey on this last point. Decision makers should not assign credal probabilities to the acts which are available to them.

Spohn notes that such probabilities are irrelevant to the ranking of the options to be used in identifying optimal or admissible options. This correct observation does not entail a prohibition against the deliberating agent assigning credal probabilities to hypotheses predicting his decision. But it serves notice on those who insist that such probability assignments may be made and, perhaps, ought to be made that they should identify some function such probability assignments can serve other than in guiding a choice among the options under consideration. Spohn reminds us that assigning such probabilities is not crucial in applications of the prescription to maximize expected utility or, for that matter, to my preferred recommendation that choice be restricted to *E*-admissible options (Levi, 1974, 1980).

Spohn offers a second more interesting argument against the agent assigning probabilities to his acts. Echoing a difficulty raised by M. Balch (1974, p. 79) against the theory of conditional expected utility proposed by Luce and Krantz (1971), Spohn notes that subjective probabilities can be elicited by offering bets in order to identify odds at which such bets are, by the agent's lights, "fair". Spohn argues:

> The readiness to accept a bet on an act does not depend on the betting odds but only on his gain. If the gain is high enough to put this act on the top of his preference order of acts, he will accept it, if not, not. The stake of the agent is of no relevance whatsoever. (Spohn, 1977, p. 115)

As I understand Spohn's argument, if an act is optimal, the agent will accept a bet on the hypothesis that he will perform the act at any betting rate no greater than 1. Although Spohn makes no mention of this point, it seems to follow that if an option is inadmissible in a decision problem, the credal probability as revealed by bets on what the agent will choose should be 0.[5]

5 To see this, suppose that the original decision problem is one where X faces a choice between options **a** and **b.** Prior to making a choice, X is offered a bet on the proposi-

This result renders principles of rational choice vacuous in applications by deliberating agents to the evaluation of the options available to them prior to choice. To avoid this untoward result, a deliberating agent ought to be prohibited from assigning credal probabilities to predictions about what the agent will choose.

The aim of a normative theory of rational choice is to provide criteria for identifying a set of options which are optimal or, at least, "admissible" in the sense that they are not ruled out by the principles of choice given his beliefs and his values. If the deliberating agent is to reach a stage in deliberation where he can use such principles to identify a set of admissible options, he must identify a set A of available options or option-descriptions from which the set $C(A)$ of admissible options is selected. No option can be admissible unless it is available. If the agent changes his mind as to what is available, then the set of admissible options can change even if everything else relevant remains the same.

If agent X recognizes **a** as an available option (description), X is certain of the following: (a) that he has the ability to choose that **a** be true given that he deliberates at time *t;* (b) that he is deliberating at time *t;* and (c) that if he chooses that a be true on the deliberation at time *t,* **a** is true. (If this condition were violated, choosing that **a** be true would be inefficacious and **a**'s truth would not be under X's control.)

There is a fourth availability condition (d) stipulating claims about which X should not be certain.[6] If X is certain that he lacks the ability to

tion that he will choose **a** where he wins a positive number of utiles W if he does choose **a** and loses a positive amount L if he chooses **b.** Now X's decision problem has been altered. He has four options - not two. He can choose **a** and accept the bet, choose **a** and reject the bet and he has two similar options through choosing **b.**

It is clear immediately that X should not choose **a** and refuse the bet or choose **b** and accept it. Choosing **a** and accepting the bet dominates the first alternative and choosing **b** and rejecting the bet dominates the second. If we assume (as could be true and is the simplest case to discuss) that the utility of one of these conjunctive propositions is the sum of the utilities of the conjuncts, which of the remaining two options X should choose depends on $W+E(\mathbf{a})$ where $E(\mathbf{a})$ is the expected value of **a** as compared with $E(\mathbf{b})-L$.

Under these conditions, as long as $E(\mathbf{a}) > E(\mathbf{b})$, X should be prepared to accept the bet even when $W=0$ so that the "fair betting rate" determining X's degree of belief that he will choose **a** should be 1. By similar reasoning, if $E(\mathbf{a}) < E(\mathbf{b})$, the betting rate for the prediction that X chooses **b** should be 1. Thus, when one of the two options is uniquely admissible, the agent is sure he will choose it. More generally the agent should be certain that he will choose an admissible option - i.e., choose rationally. The argument can be extended to show that "certain" means "absolutely certain" and not merely "almost certain".

6 The availability conditions (a)-(d) do not explicate the concept of feasibility or availability of options. They furnish some conditions on X's state of full belief or conviction which should be satisfied if X recognizes elements of a set A as options available to him. Y may disagree with X's representation of the situation. X himself may change his

choose that **a** be true given that he deliberates at time *t* and given also that his deliberation meets some other conditions *C*, X must not be certain that his deliberation meets the conditions *C*. If he were certain, then he would also be certain that he does not choose **a**. This is so even though he satisfies conditions (a), (b) and (c).[7]

In addition to identifying a set *A* of available options and, hence, of making assumptions satisfying the availability conditions (a)–(d), in order to identify an admissible subset *C*(*A*), X must know enough about his values and his beliefs (both full beliefs and probability judgements) and have enough logical omniscience and computational capacity to use his principles of choice to determine the set *C*(*A*). That is to say, his state of full belief must meet a self knowledge condition and a logical omniscience condition. We do not need perfect self knowledge or perfect logical omniscience but just enough so that if X has identified a set *A* of available options, he uses his principles of choice to identify the admissible set.

Suppose X's cognitive state meets the availability, self knowledge and logical omniscience conditions. In that case, he is certain what the elements of the admissible set *C*(*A*) are.

Suppose, in addition, that X's cognitive state meets an additional condition of smugness about rational virtue. This smugness condition states that X is certain that on the deliberation at time *t*, X will choose an admissible option.

Assuming that X's logical omniscience enables him to identify elementary logical consequences of adding this assumption to those satisfying the availability, self knowledge and logical omniscience conditions, X must be certain that he will not choose an inadmissible option.

By the availability conditions (c) and (d), it follows that no inadmissible option is an available one. The set of admissible options must coincide with the set of available ones. $C(A) = A$. Though this result is not contradictory, it implies the vacuity of deploying principles of rational choice. Such principles are supposed to offer criteria for reducing a set of options

mind about what was available to him at *t* at some other time. The conditions characterize X's state of full belief at time *t* and do not comment on whether his state does or does not contain error.

7 If the agent is certain he will choose an admissible option from among the available ones, the inadmissible options must be unavailable from his point of view. This is not to deny that the agent is certain that he is able to choose the option through his deliberation. But that conviction is consistent with his full belief that he is not able to do so through his deliberation subject to the constraint. If the agent is also certain that his deliberation is subject to the constraint, the option is not available in the relevant sense.

recognized to be feasible to a set of admissible options. Sometimes such criteria fail to furnish a reduction. If they always fail, one might as well drop the criteria. For this reason, I reject the equanimity with which F. Schick (1979) has accepted this untoward result.

Thus, at least one of the assumptions which leads to the trivialization of principles of rational choice as criteria for self criticism needs to be abandoned if trivialization is to be avoided. We cannot abandon the availability conditions, the omniscience conditions or the self knowledge condition without precluding the use of the principles of rational choice by the deliberating agent to identify the admissible options given his state of belief.[8] For this purpose, however, we can abandon the smugness about rational virtue. X should not be sure that he will choose an admissible option (i.e., rationally by his own lights) in the current deliberation (Levi, 1986a, ch. 4).

Just as a coin is capable of landing heads on a toss while at the same time it is incapable of landing heads on a toss which situates the coin in a mechanical state constraining its trajectory so that it lands tails, so too an agent may be able to perform some action through deliberation but not be able to perform that action through deliberation of some more specific kind. In the case of the coin, we may say that the coin has a 50% chance of landing heads on a toss but has a 0% chance of landing heads on a toss which situates the coin in the tails inducing mechanical state. When we consider a bet on the outcome of a toss, if we are certain that the coin is tossed, we may assign an 0.5 belief probability to the hypothesis that the coin lands heads provided we are ignorant as to whether the coin is so situated. But if we have extra information about the particular toss in question so that we are certain that the toss is of the tails inducing variety, we should assign 0 belief probability to that hypothesis. The agent may remain certain that the coin has an 0.5 chance of landing heads on a toss. But the information available to him should preclude his basing his subjective probability assessment on that chance. A similar observation applies to the deliberating agent. If the inquiring agent is certain that the deliberation in which he is engaged is one where he will not implement a given policy, then even though he knows that he is able to implement that policy through deliberation, given that he also is certain that he is not able to do so in a deliberation which terminates with his not choosing that alternative, he must base his judgements as to what is feasible to him on his information as to what he is able to do in deliberations where the option in question is not implemented. See Levi (1984, 4.1 and 4.2) for an elaboration of this argument. Background discussion of relevant conceptions of possibility and ability may be found in Levi (1977, 1979, 1980, chs. 1, 11, 12).

8 Perhaps, the agent need not be sure that if he chooses A, his choice will be implemented. That won't do because then his option is not choosing A but choosing to try to realize A, and the inquirer will presuppose that he is efficacious in choosing to try.

Perhaps, we should think of the predicament of the agent prior to his identifying his values and beliefs and making the calculations and deductions from his data requisite to determining which options are admissible. It may be argued that until the deliberation is brought to a successful fruition, the agent is not sure which options are admissible and which are not and, hence, trivialization is avoided. Even so, prior to identifying the data requisite to applying the criteria of choice and prior to making the calculations required, the deliberating agent has not applied his criteria of choice to determine an admissible set of options. To avoid trivialization and retain applicability,

I am not suggesting that the agent should be certain that he will choose an inadmissible option. Were X of this opinion, no admissible option would be recognized to be feasible, and this result is at least as objectionable as the previous one.

In footnote 5, Spohn's interesting but enigmatic second objection to allowing deliberating agents to assign credal probabilities to their own acts was elaborated into an argument showing that if they do so, the deliberating agent must be certain that he or she will choose an admissible option – i.e., choose rationally. I have just now explained why this implication leads to the vacuous applicability of principles of rational choice by the deliberating agent to his or her own decision problem. Spohn's thesis that the decision maker should not assign either determinate or indeterminate credal probabilities to hypotheses as to how he will choose is entirely justified.

All of this is predicated on the assumption that agent X is concerned with the degree of credal probability X should or should not have in deliberating as to what to do. In some contexts, X might be concerned to predict how X will act without deliberating as to what to do. When he does so, he is not functioning as a deliberating agent concerned to identify which of these options he is not rationally prohibited from making. Oblomov-like, he takes the posture of a spectator concerning his own performances. To the extent that he does so, he ceases to be an autonomous agent.

Although Spohn (1977, p. 115) seems to think otherwise, these considerations do not prevent X from taking a predictive or explanatory attitude towards his own choices at times other than the time at which deliberation is taking place any more than it precludes him from predicting or explaining the choices of others.

Thus, in the sequential choice problem represented by the extensive form of Figure 4.1, X at the initial stage might very well assign credal probabilities concerning how he will choose at stage 2 if he chooses A at stage 1 or if he chooses B at stage 2. But at stage 1, X does not regard his options at stage 2 as available to him at stage 1. His choice at stage 1 is between A and B.

In this respect, there is a significant difference between the sequential decision problem represented in Figure 4.1 and the decision problem cast in normal form in Figure 4.2 when this is understood not as a

we need to be able to apply the criteria when all the data are in and the computations made necessary to apply the principles and, given performance of this task, we need to be able to distinguish the admissible set from the feasible set relative to the information then available.

skimpier description of the same sequential decision problem but as a nonsequential, single choice node decision problem. According to the Figure 4.2 representation, X has four options available at stage 1 and none at stage 2. The question of predicting how he will choose at stage 2 given, let us say, that he chose A1 at stage 1 does not arise. Given that he chose A1 at stage 1, it is certain that he will implement A1. Perhaps, he will implement only A at stage 1 and the implementation of 1 will be delayed to stage 2. But from X's perspective at stage 1, it is certain that 1 will be implemented later given the choice of A1 at stage 1. If X did not judge choosing A1 to be efficacious, he would not regard choosing A1 as an option available to him. From his stage 1 perspective, his future self has no say in the matter.

The situation is different when X's predicament is represented by Figure 4.1. In that case, at stage 1, X has control over whether A or B is true but not over whether A1, A2, B3 or B4 is true. He can, however, make a prediction as to how he will choose at stage 2 and guide his choice of A or B accordingly.

So agent X may coherently assign unconditional credal probabilities to hypotheses as to what he will do when some future opportunity for choice arises. Such probability judgements can have no meaningful role, however, when the opportunity for choice becomes the current one. Indeed, whenever X regards his acts as currently subject to X's control, X cannot coherently assign unconditional probabilities to hypotheses as to what X will then do. Deliberation crowds out prediction.[9]

Someone may object that by refusing to take into account X's views about whether he will choose rationally, relevant information is suppressed. That, however, is not so. The only information X is entitled to suppress and, indeed, should suppress as a deliberating agent concerns whether X will choose rationally. X should not be certain that he will choose in conformity with his principles of rational choice and he should not be certain that he will violate them. He should make no assumptions as to the chances or statistical probabilities of his choosing rationally or irrationally. Descriptions of X as choosing rationally or irrationally in the decision problem under consideration are irrelevant information in the context of X's deliberation.

9 The issue of foreknowledge of one's choices is discussed by Shackle (1969), Jeffrey (1977) and Schick (1979). The suggestion that one should not be certain that one will choose an admissible option is proposed in Levi (1986a). T. Seidenfeld in discussion and W. Spohn in print prompted me to address the question of the relevance of probabilities of predictions of choice to determining betting rates and, hence, to link the question of probabilities of options with the response to the foreknowledge conundrum advocated in Levi (1986a).

Nonetheless, even if X removes hypotheses about the rationality of his current choices from the algebra of propositions to which he assigns probabilities, he may still have information in his corpus K of certainties, evidence or knowledge (the set of propositions X takes for granted at time *t*) which entails that he will or will not choose a given option or which assigns a chance or statistical probability to his choosing (or not choosing) a given option. Such information obligates X via direct inference to assign credal probability to his choosing (not choosing) that option. As such, it precludes X's considering that option as one which is available in the context of deliberation.

As long as X does not have any knowledge entailing that he will (will not) choose true a given act-description or specifying chances of his choosing (not choosing) true such an act-description, he is not committed to assigning credal probabilities to hypotheses as to what he will choose and as a deliberating agent should not do so.

This view has profound ramifications for many topics. For example, when it is assumed that it is common knowledge among participants in a game that each player will play rationally, player X is assumed to be certain that he will choose rationally or assigns probabilities to his doing so. On the view I am advocating, this is incompatible with his having available to him options other than admissible ones - the implication which is fatal to decision theory in general and game theory in particular at least when these theories are used prescriptively.[10]

Of more immediate concern, however, is the relevance of all of this to consequentialism and sequential choice. In this connection, the first point I am belaboring is that (pace Jeffrey) Savage is surely right in

10 Pettit and Sugden (1989) take notice of the fact that common knowledge conditions in game theory imply that the options available to players coincide with the options admissible for them, but they distinguish that case from the case of common belief where they do not recognize such an implication. As Schick rightly notes, however, the puzzle we are attending to is a puzzle about forebelief as much as about foreknowledge. The options available to the agent in a decision problem are the options the decision maker recognizes as available to him. If the agent is certain he will not choose a given course of action even though he is certain that he is able to do so through deliberation and his conviction is correct, that option is not available to him. Thus, if the option dominates alternatives which are available to him, the injunction against choosing a dominated option is not operative; for it applies (whenever it does) only when the dominated option is dominated by an available option. The notion of availability relevant in discussing criteria for rational choice intended for agents in deliberation is a notion of feasibility as recognized by the deliberating agent and not by a third party such as a court of law or God. It is entertainable that other notions of availability are appropriate when assessing moral or legal responsibility. Relative to such notions, the distinction between common knowledge and common belief might be relevant. I deny it is relevant in either game theory or decision theory.

contrasting act-descriptions with consequence-descriptions and state-descriptions. Act-descriptions, insofar as by this we mean propositions representing available options, are not to be assigned unconditional probabilities. Hence, any proposition whose truth is probabilistically dependent on which option is exercised is not to be assigned unconditional probability. Moreover, no proposition whose truth is probabilistically independent of the option chosen should be assigned unconditional utility unless every available option bears the same value when that proposition is true.

More crucial to the concerns of this essay, moreover, the view of availability just sketched suggests that the options available to the agent at the initial node in a sequential decision problem in the extensive form of the sort illustrated by Figure 4.1 do not correspond to the options available in the nonsequential normal form version illustrated by Figure 4.2.

Advocates of the reduction of extensive to normal form might concede this point and still insist that to be rational the agent should follow the path in the Figure 4.1 predicament corresponding to the option he should choose in the Figure 4.2 predicament. But the argument for this view cannot rest on the assumption that precommitment to following any one of the paths in the Figure 4.1 decision tree is an available option to the decision maker at the initial node. The question before us is whether a version of consequentialism presupposed by "modern normative decision" theory can be deployed to make an alternative case for the conclusion.

To explore this matter further, we shall turn our attention to consequence-descriptions, state-descriptions and consequentialism.

3. WEAK AND STRICT CONSEQUENTIALISM

Loosely speaking, consequentialism as it is relevant to "modern normative decision theory" holds that the evaluation of options by a decision maker in a given context of choice should be determined by the values the agent assigns to the consequences of the several options and the probabilities of these consequences being realized given the implementation of these options. Different kinds of consequentialism may be distinguished, however, depending on how we understand what is to count as a consequence of an option. I have already argued that the set A of propositions representing options available to the decision maker depends on the agent's corpus $\underline{K}$ of certainties or full beliefs. Given $\underline{K}$ and any option description a_i in A, let O_i be a set of propositions such that $\underline{K}$

and a_i ($\underline{K}+a_i$) entail that exactly one element of O_i is true and such that each element of O_i is consistent with $\underline{K}+a_i$. Let O be the union of the O_i's.

Any representation of X's decision problem in terms of the set A and, for each a_i in A, a set O_i such that each element of O is assigned an unconditional utility is a ***weak consequentialist*** representation and the elements of O_i are consequence-descriptions in the weak sense of a_i.[11]

Every decision problem is representable in weak consequentialist form; for every available option bears an unconditional utility and by identifying the elements of O with the elements of A a weakly consequentialist representation is available.

A representation of X's decision problem is ***nontrivially weak consequentialist*** if and only if for some a_i, a_i is not the sole element of O_i. The expected utility $E(a_i)$ of a_i is $\Sigma p(o_{ij}/a_i)u(o_{ij})$.

Observe that if there is a nontrivial weak consequentialist representation of a given decision problem, then for every a_i in A, the unconditional utility $u(a_i)$ should equal $E(a_i)$ and this should be so for every nontrivial weakly consequentialist representation.

For every act-description a_i in A, one can always construct a set O_i by taking any set Q_i of sentences such that $\underline{K}+a_i$ entails exactly one element

11 This formulation needs some modification in order to be adequate to our purpose. As it stands, it presupposes that the consequence-descriptions in O are assigned numerical utilities unique up to a positive affine transformation. Our aim in this discussion is to explore the extent to which consequentialism entails that available options in A be weakly ordered with respect to value when sequential choice is taken into account. If we characterize consequentialism as weak consequentialism - i.e., the thesis that every decision problem must be representable in weak consequentialist form - there is no need to invoke considerations of sequential choice to yield the result. Assuming that the expected utility of an option is its utility, the formulation we have given ensures that options representable in weak consequentialist form order the available options in A without any consideration of sequential choice.

To avoid such trivialization of Hammond's thesis, let us suppose that the agent X facing a decision problem assigns value to some propositions but not to others. The propositions assigned value belong in the *value domain*. Other propositions do not. We assume that all elements of A are in the value domain and that the conjunctions (consistent with $\underline{K}$) of elements of A with any proposition are in the value domain (so that for every option-description in A every proposition has a value conditional on that option-description). In addition, if for every a_i and $a_{i'}$ in A such that $h\&a_i$ and $h\&a_{i'}$ are consistent with $\underline{K}$, $h\&a_i$ is equivalued with $h\&a_{i'}$, h is in the value domain. No other propositions are in the value domain. The evaluation of the propositions in the value domain need not be representable by a utility function unique up to a positive affine transformation over the value domain. The evaluation may yield at most a weak ordering of the value domain. Perhaps only a quasi ordering is obtainable. In any case, when we say that a proposition is assigned unconditional utility, we mean only that it is in the value domain. If $h\&e$ is in the value domain and a utility function is defined over the value domain, the utility of h given e is equal to the utility of $h\&e$.

of Q_i and each element is consistent with $\underline{K}+a_i$ and by letting O_i be the set of sentences of the form $a_i \& q_{ij} = o_{ij}$ where the o_{ij}'s are assigned unconditional utility. This representation is the weak consequentialist representation generated by the set of descriptions Q. Every decision problem, therefore, is representable in a nontrivial weakly consequentialist form as long as the conceptual framework is sufficiently rich to allow for the construction of sets Q_i.

Weak consequentialism (the idea that every decision problem should be representable in weak consequentialist form so that the evaluation of available option-descriptions is determined by the evaluation of the consequence descriptions and the probabilities of their occurrence given the implementation of the options) ought not to be a focus of controversy. Critics of consequentialism in moral theory cannot be complaining about weak consequentialism since it is sufficiently flexible to accommodate any mode of nonconsequentialist evaluation they care to consider. Moreover, strict Bayesians advocating the principle of expected utility maximization need not presuppose any stronger version of consequentialism than weak consequentialism.

Is weak consequentialism so understood what Hammond has in mind? Hammond writes:

> As a normative principle, however, consequentialism requires everything which should be allowed to affect decisions to count as a relevant consequence - behaviour is evaluated by its consequences, and nothing else. (1988, p. 26)

Hammond seems to be saying that whatever is an object of value in the context of deliberation qualifies as a consequence. And, moreover, he seems to think that such consequentialism is neutral with respect to disputes over "practical" (substantive?) normative principles. These demands are met by weak consequentialism. Weak consequentialism does not state in advance which descriptions are in the domain of consequence descriptions. That is a matter for practical or substantive values - as Hammond says.[12]

12 I would supplement Hammond's observation by pointing out that the value domain of consequence descriptions for agent X (see footnote 8) is relative to (i) the agent's corpus $\underline{K}$, (ii) the agent's values and (iii) the agent's credal probability judgements. Thus, whether *h*&*a* has the same value for every option *a* in *A* and, hence, is in the value domain depends on factors (i) and (iii) as well as factor (ii). *h* could be equivalent given K to a set of exclusive and exhaustive hypotheses. It may meet the requirements for belonging in the value domain in virtue of not only the agent's values but the corpus and the credal probabilities over these alternatives. Still I agree with Hammond that consequentialism imposes a constraint on consequence descriptions. Which systems of descriptions meet the constraints is a substantive issue.

The neutrality of weak consequentialism is greater, however, than Hammond may be prepared to require of consequentialism. For any decision problem whatsoever, there is at least one weak consequentialist representation. No one can avoid being a weak consequentialist - not even Kant and Bernard Williams. I am not sure that Hammond intends his version of consequentialism to be as weak as all that.

It is not clear, for example, whether Hammond means to allow option-descriptions to be consequence-descriptions as well. Nor is it obvious that he means to allow as consequence-descriptions, weak consequence-descriptions which are conjunctions of option-descriptions with other sentences. In any case, Savage seems to have endorsed versions of consequentialism imposing stronger demands than weak consequentialism entails.

A representation of a decision problem is in *strict consequentialist* form if and only if (i) it is nontrivially weak consequentialist and (ii) it is generated by a set of descriptions Q in the value domain where for every pair of options a_i and $a_{i'}$, if q_{ij} is the same description (sentence, proposition) as $q_{i'j'}$, $a_i \& q_{ij}$ is equivalued with $a_{i'} \& q_{i'j'}$. In this case, the elements of Q are *strict consequence descriptions.*

Strict consequentialism is the view that for every decision problem there should be a representation in strict consequentialist form. Unlike weak consequence descriptions, strict consequence descriptions never have act-descriptions as conjuncts. In such representations, there is a sharp separation between act-descriptions and consequence-descriptions. The value of a consequence-description does not depend upon which act determines it to be true. What "affects" decisions are the values of the strict consequence-descriptions. Act-descriptions have value and, indeed, unconditional value, but their unconditional value is supposed to be "dependent" on the unconditional value of their strict consequence-descriptions. At the level of generality at which this discussion is being carried on, the notion of "dependence" cannot be explained precisely because such an explanation will appeal to principles of rational choice determining which options are admissible from a set of available options. We have been avoiding the assumption of such principles. However, if one is an advocate of the expected utility principle, strict consequentialism may be construed as saying that the value of an option-description must be represented as an expected utility derived from information about the utilities of strict consequence-descriptions and probabilities assigned to them conditional on the option-description being true.

Savage was committed to some form of strict consequentialism. This is a far more substantive thesis than weak consequentialism and, indeed, has value implications which I find far too restrictive (although Savage seemed to think that the restrictions could be finessed somehow).

Strict consequentialism does not imply that strict consequence-descriptions represent possible effects of the events described by act descriptions. To the contrary, there is nothing to prevent a consequence description, when true, from describing the same event as an act-description. If George has forgotten whether he borrowed the $10 from Ron or from Nancy but is sure he borrowed from one of them, he may face a choice of paying Ron or paying Nancy. Given his corpus, it is a serious possibility that if George pays Ron the $10, the event description "the paying of his debt" refers to the same event as "paying Ron the $10". Causality on almost nobody's account enters into the matter. Yet, "George pays his debt" is a possibly true consequence-description for the act-description "George pays Ron $10".

We could envisage a version of consequentialism which, in addition to requiring representation of decision problems in strict consequentialist form, insisted that the applicability of strict consequence-descriptions be causally dependent on the option chosen. Fortunately, we do not have to worry about this version of consequentialism or the vexing problems of clarifying the notion of causal dependency it deploys. Savage does not seem to have presupposed such causal versions of consequentialism. Nor, as far as I can see, do other students of decision theory including advocates of so-called causal decision theories.

Savage's motive for adopting a stronger version of consequentialism is his interest in deriving an agent's belief probabilities and utilities from information about his preferences for the "acts" or options he faces or, more accurately, for his preferences for a hypothetically available set of options in which the options the agent actually faces are embedded. The axioms he imposed on such preferences presupposed that the hypothetically available set of options could be represented in strict consequentialist form and not merely in weak consequentialist form.

To achieve his purpose, Savage endorsed a still stronger form of consequentialism. As Hammond pointed out in the passage cited initially, Savage required that options be representable in what I shall call "state functional form" so that each option-description is equivalent to a representation of an option as a function from a system of "state-descriptions" to a system of strict consequence-descriptions.[13]

13 Causal decision theorists insist that issues of causal dependence are relevant to determining which options are admissible; however, they do not insist that there be at least one causal and strict consequentialist representation for every decision problem.

As just noted, the chief motivation for the additional restrictions imposed by Savage appear to be that they facilitate derivation of numerical probabilities and utilities from preferences over act-descriptions when suitable axioms are adopted constraining the preferences. But we ought not to impose conditions which support such derivation as requirements of rationality merely for this reason. I would regard strict consequentialism as excessively restrictive as a condition of rationality. A fortiori, state functional, strict consequentialism ought not to be accorded that status.[14]

14 Given X's corpus $\underline{K}$, *S* is a set of *protostate descriptions* if and only if (i) $\underline{K}$ entails that exactly one element of *S* is true and each element is consistent with $\underline{K}$ and (ii) the elements of *S* are not in the value domain.

Protostate-descriptions, like state-descriptions, are not objects of unconditional utility evaluation. We have not required, however, that protostate-descriptions be assignable credal probabilities not conditioned on the option chosen. That is to say, we have not required protostates to be probabilistically independent of the option chosen. In this respect, they resemble consequence-descriptions.

Given X's corpus $\underline{K}$ and state of credal probability judgement *S* is a set of *state-descriptions* if and only if (i) *S* is a set of protostate-descriptions relative to $\underline{K}$ and (ii) for every option description **a** in *A*, the state of credal probability judgement for *S* given **a** is the same.

Savage focused attention on decision problems representable in a form which is at once strictly consequentialist and where there is a set *S* of state-descriptions such that for every option a_i of *A* and every element s_j of *S*, there is an element q_{ij} of *Q* such that $\underline{K} + a_i \& s_j$ entails q_{ij}. When this is the case, the corpus $\underline{K}$ and option-description a_i determine a function from the *S* to *Q*. Such a representation is *state functional and strict consequentialist.* If *S* is a set of protostates but not states, the representation is *protostate functional.*

Two or more option-descriptions may determine the same function from protostate-descriptions to consequence-descriptions. Moreover, in general two such option-descriptions need not be equivalued. Savage, however, insisted that if two acts have the same consequences in every state of the world, "there would be no point in considering them different acts at all" (Savage, 1954, p. 14). That is to say, they would be equivalued. This is a very substantial value and probability assumption. For one thing, it presupposes strict consequentialism. For another, if the value of an option is supposed to be its expected utility, a protostate functional, strict consequentialist representation presupposes that the probability distribution over the protostates is independent of which of the acts yielding the same consequences in the same states is chosen. This is not quite probabilistic independence of states from acts. It does not presuppose that the probability distribution over *S* is the same for all elements of *A* but only for those yielding the same consequences for the same states. Savage, who is not merely protostate functionalist but state functionalist, endorses axioms which ensure this probabilistic independence. It is useful to keep in mind, however, that assumptions about probability and utility are built into the requirements of even protostate functionalist consequentialism which are not explicit in the axioms proposed by Savage or in alternative systems.

Savage and many others who are state functional, strict consequentialists claim that acts are or are represented by functions from states to consequences. I do not deny that when decision problems are representable in state functional, strict consequentialist form, one can replace the option- or act-descriptions by characterizations as

Students of the foundations of Bayesian decision theory as well as non-Bayesian decision theorists have registered reservations with the state functional, strict consequentialism of Savage and have sought to mitigate the severity of its demands. Whatever tinkering they do, however, seems to leave unsullied the strict consequentialist requirement. (See, for example, Fishburn (1981).) The thesis that "acts are valued by their consequences", which is "almost unquestioned" by "modern normative decision theory", seems to me to be best captured by strict consequentialism. Weak consequentialism is, indeed, unquestioned and rightly so. But students of decision theory have tended to assume that the stronger strict consequentialist condition is in place in their work. I fail to see that anything relevantly stronger than strict consequentialism has been "almost unquestioned". And even critics of utilitarianism and utilitarian-like moral theories may be willing to concede that in some contexts the conditions for representing a decision problem in strict consequentialist and, indeed, even in strict consequentialist, state functional form may be satisfied. Hence, in spite of its excessively restricted character, it will be useful to compare the assumptions of strict consequentialism with the version of consequentialism Hammond investigates.

functions from states to consequences. However, in some situations it may be important to consider that there are two or more options which determine the same function. As long as the corpus, the state of credal probability judgement and the state of utility judgement remain what they are, the several options are equivalued. But such judgements could change. As a consequence, the representation of the decision problem could cease being state functional or even strict consequentialist. There is nothing in the Savage axioms which make explicit the properties of the corpus, state of credal probability judgement and the state of value judgement which secure the state functional representability of the elements of *A*.

Given a decision problem representable in strict consequentialist form, is it always possible to obtain another representation, which is equivalent given the corpus $\underline{K}$, in protostate functional, strict consequentialist form? If so, is it possible to obtain a representation which is in state functional, strict consequentialist form?

Unless one invokes the use of conditionals (as is tacit in Fishburn (1970) and explicit in Gibbard and Harper (1978)), or, barring that, introduces other enlarged conceptual resources, the answer to the first question seems to be negative. And as Gibbard and Harper appreciate, even if we indulge in the use of conditionals to construct protostates, we cannot guarantee that the protostates will qualify as states.

I do not think that conditionals ought to be treated as truth value bearing propositions (Levi, 1977, 1979, 1988, 1996). For this reason, I agree with Spohn (1977) that neither protostates nor states should be constructed as conjunctions of conditionals with elements of *A* in the "if" clause and elements of *Q* in the "then" clause. I am doubtful of the availability of acceptable modes of conceptual innovation which can be deployed to obtain the desired representations. Just as strict consequentialism is stronger than weak consequentialism, protostate functional strict consequentialism is more restrictive than strict consequentialism, which, in turn, is weaker than state functional, strict consequentialism.

4. HAMMOND'S CONSEQUENTIALISM

Hammond introduces a conception of consequentialism and claims that it does entail ordinality. Consequentialism as he understands it is a constraint on "behavior norms". Our next task, therefore, is to explain the notion of a behavior norm and to state what a consequentialist behavior norm amounts to. With this understood, we will then be in a position to state the relation between Hammond's consequentialism, strict consequentialism and ordinality.

For the present purpose, we can focus on decision problems in extensive form, where every node except a final node is a decision node. There are finitely many terminal nodes each of which is represented by propositions in the value domain Y generated by a set Q of consequence descriptions.[15] We are assuming that the elements of Y are strict consequence-descriptions as well as "terminal" in the sense indicated. Hence, the value of any option available to a decision maker at a decision node in the decision tree is determined by the values of the elements of Y and the probability of realizing such elements conditional on choosing the option.

Consider any such decision tree. A *behavior norm* specifies for any such tree, and any decision node in the tree, the admissible set of options (i.e., "branches") from the set available at that node. It may be represented by a function $\beta(T,n)$ where T is a decision tree and n is a node in the tree. Given any such tree T, let $T(n)$ be that subtree obtained by considering the paths emanating from node n in T. $T(n)$ is a tree and, hence, the behavior norm should be defined when it is an argument. In particular, $\beta(T(n),n)$ should have a value. Hammond insists that in order for the behavior norm to be an "accurate description" of behavior, $\beta(T(n),n)$ should equal $\beta(T,n)$. Since, by hypothesis, there are no features of the situation relevant to evaluating value other than those packed into the consequence descriptions at the terminal nodes, if the agent in tree T and at node n recognizes a certain subset of available options as admissible, he should do the same even if the tree began at node n as long as he is wedded to the same behavior norm. This seems to be Hammond's position.

15 The elements of Y may belong to the set Q, they may be descriptions of lotteries on finitely many consequence descriptions in Q or they may be functions from protostate descriptions to consequence descriptions in Q (or lottery descriptions on Q) relative to the corpus at the last choice node and the information as to which option is chosen at that node. Whether we assume protostate functional or state functional consequentialism, this must be possible.

Now Hammond claims that if the agent chooses in conformity with a given set of principles of rational choice (whatever they may be) and is consequentialist, his behavior norm for all trees in the domain will meet a certain condition which I shall now explain.

Let $F(T,n)$ be the consequence descriptions which are terminal nodes of the subtree $T(n)$ of T. $C_\beta[F(T,n)]$ is the set of consequence-descriptions the choice of which the agent would regard as admissible were the agent to have available to him at a single choice node the choice between the elements of $F(T,n)$. The subscript β serves to remind us that a single shot choice between consequence descriptions is itself a tree with a single decision node, that the behavior norm is defined for that tree and that the "choice function" C_β is the one determined by the given behavior norm.

Suppose then the agent is at node n in tree T or at node n in subtree $T(n)$. In general, the agent will not have available at that stage a direct choice between consequences but the behavior norm will specify the set of admissible options $\beta(T,n)$ at that node and, assuming that the agent obeys the behavior norm at every subsequent stage in the tree leading from an admissible option, will determine a set of admissible subsequent paths in the decision tree. The set of consequence-descriptions which are terminal nodes in this set of admissible, subsequent paths is the set $\Phi_\beta(T,n)$. In effect, taking n to be the initial node in subtree $T(n)$, $\Phi_\beta(T,n)$ is the set of possible consequences of following the behavior norm in the agent's extensive form sequential decision problem, and $C_\beta[F(T,n)]$ is the set of possible consequences of following the behavior norm in the corresponding nonsequential strategic or normal form decision problem when the agent's evaluation of the consequences is held fixed.

According to Hammond's version of consequentialism, the set of admissible option sequences in the extensive form, sequential decision problem should have as terminal nodes the same set of consequences as the corresponding nonsequential, normal form decision problem.

Hammond's consequentialism

$$\Phi_\beta(T,n) = C_\beta[F(T,n)]$$

Return to the schematic example of an extensive form decision problem represented by Figure 4.1 and the corresponding normal form decision problem represented by Figure 4.2. If the agent prefers c_1 over c_2 over c_3 over c_4, then it is clear that as long as the Figure 4.2 predicament is in strict consequentialist form (i.e., as long as the c_i's are strict consequence descriptions), the agent should choose A1. In the Figure 4.1

predicament, at the second stage, if the agent had already chosen A, he should choose option 1 and if he had already chosen B, he should choose 3. Hence, at the first stage, he should choose option A; for in that event, if he follows the behavior norm, he will end up with c_1. Thus, under the hypothesis that the consequences are weakly ordered, the verdicts of strict consequentialism and Hammond's consequentialism coincide.

Consider, however, Hammond's well known treatment of the potential addict which he discusses afresh in his 1988 paper. The potential addict faces a choice at the initial time t_1 whether to try out some addictive substance. He does not want to become permanantly damaged by addiction but to enjoy briefly the pleasures of the substance without damage. So he prefers pleasure without damage to complete abstinence to damage. If he chooses to partake of the substance, at t_2 he has a choice of refraining from further use, with the result of pleasure without damage, or continuing, with the result of permanant addiction and damage. We suppose that at t_2, he has already become addicted. The decision tree is represented in Figure 4.3.

Consider, at least as a thought experiment, the corresponding single choice predicament where the potential addict has three options: pleasure without damage, complete abstinence and permanant damage. Given the preferences we identified, we suppose that the behavior norm specifies a uniquely admissible option for this case: Pleasure without damage. This is the value for the choice function C_β for this case. The only kind of behavior norm covering the extensive form problem as well which meets Hammond's consequentialist requirement is the one which recommends partaking at stage 1 and stopping at stage 2. By hypothesis, however, the potential addict is incapable of conforming to the behavior norm. He might even know this at the initial choice node. What behavior

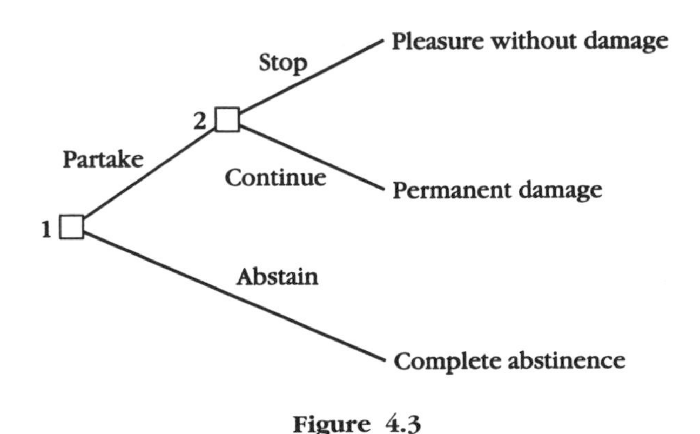

Figure 4.3

Table 4.1

	Tree	Node	Admissible set
Naive β_1	T(1)	1	Partake
	T(1)	2	Continue
	T(2)	2	Continue
Sophisticated β_2	T(1)	1	Abstain
	T(1)	2	Continue
	T(2)	2	Continue

norm should he live by, compatible with the behavior norm specified for the normal form problem? Hammond singles out two candidates, depicted in Table 4.1.

Both of these behavior norms are consistent in Hammond's sense. Neither of them satisfies the condition of Hammond's consequentialism, as Hammond himself emphasizes.

The naive behavior norm violates the precepts of strict consequentialism. At stage 1, the agent knows that assuming he partakes, he will end up with permanant addiction. He also knows that if he does not partake he will end up with total abstention. By hypothesis, at the first stage he would prefer the latter consequence over the former. If only the consequences matter to him, then as a good strict consequentialist, he should choose to abstain. Naiveté leads to abandonment of strict consequentialism as well as Hammond's consequentialism.

In contrast to naiveté, sophistication meets the demands of strict consequentialism. However, Hammond's consequentialism is violated. Hammond's consequentialism requires that applying the behavior norm leads to consequences that would be chosen in a normal form version of the problem. If the agent had his druthers, he would have the temporary pleasure of partaking without lasting addiction and damage. So Hammond's consequentialism recommends partaking at the first stage and then stopping at the second. But, by his own lights, the agent cannot partake and then stop. The sequence of choices is not an available option. The sophisticated decision maker realizes this. His behavior norm is implementable, as is the norm of the naive decision maker. Unlike the naive decision maker, however, the sophisticated decision maker makes a decision at the initial node that yields the best consequences from those accessible to him.

Hammond's consequentialism is a much more demanding prescriptive standard than strict consequentialism. If the potential addict is certain that he will not stop at the second stage, he is certain that partaking will lead to permanant addiction. That is the strict consequence of his partaking from his point of view. Abstaining coheres with strict consequentialism.[16]

Hammond's consequentialism requires the decision maker at the initial stage to proceed as if he had available to him at the initial stage the option of obtaining any element of Y - i.e. a consequence-description at a terminal node of his decision tree. The example of the potential addict illustrates one of several ways in which this condition can break down. It

16 Perhaps, it will be objected that the potential addict is not certain that he will choose to continue at the second decision node. He may judge this highly probable but not certain.

If he does that, he is in the position of predicting how he will choose at the second stage. As we pointed out, at the first stage the agent may assign credal probabilities to hypotheses as to his future choices at later stages conditional on his initial choice.

If the agent reaches the second stage and knows it, then these hypotheses become option descriptions. Because the agent at stage 2 regards himself as having a choice at that stage, he does not convert his erstwhile conditional probabilities as to how he will choose into unconditional probabilities as good Bayesians will do. For reasons we have already explained, he should refrain from probability judgement concerning these propositions.

Consider now the hypothesis that the potential addict will reach the second stage choice node - i.e., he will face the choice between stopping and continuing. That hypothesis is a strict consequence-description from the initial point of view. Its value (as assessed relative to the situation at the initial node) is determined by the values assigned to the hypotheses that the potential addict will stop or continue and the probabilities of these hypotheses conditional on his having taken the drug at the initial stage.

Thus, the potential addict might assign a high probability at the initial stage to his continuing with the drug at the second stage, thereby sinking into abject addiction with permanant damage. His second stage opportunity for choice is an uncertain prospect with an expected utility. Even if we do not assume that uncertain prospects are evaluated with respect to expected utility but with respect to some other sort of index, if the propositions at the nodes immediately following the choice node representing the current opportunity for choice are strict consequence-descriptions, relative to the options available at the initial node, so is the description of the opportunity for choice - i.e., the choice node.

Hence, for a strict consequentialist, the uncertain prospect is itself a strict consequence-description. But it is not in the terminal set Y. However, unlike the elements of Y accessible from choice node [2], it is accessible to the potential addict at the initial node. From the initial point of view, the two strict consequence-descriptions directly under the potential addict's control are the uncertain prospect describable by the opportunity for choice at the second node and the terminal consequence of complete abstinence. Strict consequentialists would recommend that the potential addict choose an admissible element from this pair without pretending to have available directly under his control all elements of Y.

is presumed to be false in that case that the agent is able to choose the strategy of first partaking and then stopping. At the initial stage, the potential addict has control over whether he partakes or not. He has no control over what he will choose at the second stage. Hence, the consequence of partaking and then stopping is not available to him.

Suppose, for example, that our strict consequentialist potential addict were a strict Bayesian like Savage and were deciding whether to partake or abstain on the basis of expected utility. The expected utility of partaking is determined by the utilities of the elements of *Y* accessible from choice node [2] and the probabilities of the potential addict stopping or continuing to take the drug at the second stage. This procedure violates Hammond's consequentialism. But it does not betray the strict Bayesianism of Savage, Ramsey, de Finetti, et al. Strict Bayesianism may entail consequentialism in some sense or other but not in Hammond's sense.

Hammond himself admits that given the potential addict's predicament, "sophisticated behavior seems clearly the best, despite its violation of consequentialism" (1988, p. 36). Hammond immediately goes on to say:

> This does not imply, however, that consequentialism is irrational. Rather the potential addict is really two (potential) persons, before and after addiction, and the decision problem has to be analysed as a "game" between two "rational players". (1988, p. 36)

Thus, for Hammond, the fact that the best thing for the potential addict to do is to behave according to the nonconsequentialist sophisticated behavior norm does not undermine the rationality of consequentialism as he understands it. He claims to be restricting his discussion to "single person decision theory". Multiperson decision theory is to be left for future consideration. The predicament of the potential addict belongs to multiperson decision theory. Hammond does not explain why he thinks the potential addict is really two persons before and after addiction. Is it because the potential addict has different preferences or values after addiction than before or is it because the change in preferences is not the product of well reasoned reflection?

Suppose we heed the entirely sensible attitude favored by Kadane and Larkey (1982) that when an individual faces a choice in a game against another individual or set of individuals, he should regard hypotheses about the choices of the other players as objects of probability judgement. In a similar vein, at the initial choice node, the potential addict should treat hypotheses as to his future choices as objects of probability judgement. In this respect, the predicament of the potential addict does indeed resemble that of a participant in a game. The other player is the potential addict himself at the second choice node.[17]

However, in this sense, every sequential decision problem is a game between the decision maker at the initial node and the decision maker at subsequent nodes. It does not matter whether the agent has or has not altered his values at the several nodes and, if he has, whether he has done so in a reasonable fashion or in the grip of an addiction. If Hammond thinks that the predicament of the potential addict is a multiperson decision problem not covered by his discussion, then all sequential decision making is multiperson decision making. Hammond's account of consequential behavior norms then has application only to trees with single choice nodes - counter to his intention.

Hammond must have in mind a different view of multiperson decision making than the one I have just mentioned. He owes us an explanation of what it is. Regardless of the answer, Hammond's consequentialism is stronger than strict consequentialism in two respects: (a) it prohibits recognition of opportunities for choice as uncertain prospects counting as consequences and (b) it demands that a behavior norm in a decision tree yield the same terminal consequences as the corresponding normal form decision problem does.

It is now time to consider whether strict consequentialism entails ordinality. We shall address a predicament where the decision maker is to choose autonomously at several stages without changing his preferences or values from one decision node to another. Philippa has had the good fortune to be offered beginning teaching positions at three philosophy departments at universities A, B and C. Philippa judges the teaching conditions and intellectual environment at A vastly superior to B, which is in turn negligibly superior to C. C offers much better salary and tenure prospects than B, which in turn is slightly better in these respects than A. University B has never hired a woman as a faculty member. C has hired a woman but does not have one now. A has several women on its faculty. Philippa is concerned primarily with teaching conditions and intellectual environment on the one hand and salary and tenure prospects on the other. The record of the university in hiring women would matter to her only if she could not make up her mind on the basis of the primary considerations.

17 This claim implies nothing more about the addict's identity over time than that at the initial choice node, the opportunities for further choice he expects to realize through the decision he makes are uncertain prospects.

On the basis of the primary considerations, Philippa is clear that she will not choose B. It is nearly as bad as the worst of the three according to both dimensions of value she recognizes as primary. If she countenances potential resolutions of the conflict between these two dimensions, no compromise rates B on a par or better than both the other alternatives. So she rules B out as inadmissible on primary considerations. But she cannot make up her mind between A and C. In this case, however, she appeals to the desirability of the presence of a woman at university C and chooses in its favor.

Observe, however, that if her choice had been between B and C without the opportunity to choose A, professional considerations would not have enabled her to decide. B would have been better with respect to teaching conditions and intellectual environment but would have paid less and had poorer tenure prospects. As a feminist, Philippa would be better off breaking the ice at university B than being the second woman on the staff at C. So she chooses B.

This is an example of a violation of an important condition on choice functions (Sen's property α). C is uniquely admissible in the three way choice, but B is admissible and C inadmissible in the two way choice. Violation of this choice consistency principle indicates either that when facing the pair of options, Philippa has changed her values from those she endorsed when facing three options or that Philippa does not have an evaluation of her options representable as a weak ordering. As I have portrayed her predicament, her behavior is not a manifestation of a change in values but derives from the fact that her primary values are in unresolved conflict (Levi, 1986a) and cannot induce a weak ordering on her options. This portrait seems consistent and, more crucially, there is nothing in it to suggest that Philippa's values are incoherent or irrational.

Philippa's decision problem was nonsequential. She faced a choice between three options. But suppose her choice is a sequential one. Suppose university A requires a yes or no response tomorrow, whereas universities B and C call for a response within a week of tomorrow. In a three way choice, I have suggested that Philippa would choose C. If Hammond's consequentialism were to obtain, the behavior norm would have to recommend refusing university A at the first stage and choosing C at the second. But that behavior norm does not conform to the vision of her values I have suggested. If she does refuse A at the first stage, Philippa has a choice between B and C and, so I have suggested, she should choose B. If Philippa, at the first stage, is certain that she will choose B at the second stage, she sees her first stage options as choosing

A and refusing A, followed by choosing B. If her values are as I have specified, she should refuse A and choose B. However, in the normal form version of Philippa's predicament, Philippa should choose C. Hammond's consequentialist condition is clearly violated.

Philippa, I submit, has been a loyal strict consequentialist. Given her conviction that she will choose B at the second stage, the only "consequence-descriptions" accessible to her from her initial point of view are A and B. C is not accessible. The reason is that Philippa recognizes at the initial stage that should she refuse A at the outset, she will face another choice. The second choice is an autonomous one. From her initial point of view, how Philippa will choose subsequently is no more under her control at the initial stage than how someone else would choose if the decision as to whether Philippa goes to B or C when A is rejected were left in the hands of a third party. Philippa can at best predict what her second stage self will choose. From the vantage point of the first choice node, Philippa regards her second stage opportunity for choice between B and C as a strict consequence of refusing A. That is, in point of fact, the consequence accessible to her. If she is sure she will choose B, then B is an accessible consequence as well. If she is uncertain, neither B nor C is accessible. Only if she is certain that upon refusing A, C will be chosen, is C accessible. In the situation we are envisaging, C is inaccessible. This is not because at the second stage Philippa is crazed by addiction. To the contrary, it is because Philippa has nonarbitrarily and resolutely stuck by her values and principles. She is not of two minds. Her second stage self endorses the same values as her initial stage self. At both stages, Philippa is in unresolved conflict. Yet Philippa behaves like a good strict consequentialist.

Philippa's predicament illustrates why the strict consequentialism presupposed by Savage and other students of "modern normative decision theory" does not imply ordinality. As mentioned earlier, T. Seidenfeld has shown that when decision trees terminate in strict consequence-descriptions, the independence postulate should be satisfied whether or not reduction of extensive to normal form obtains. I have argued elsewhere that the principle of revising probabilities by temporal conditionalization can be justified by betting arguments along the lines of Teller (1973) only if the reduction of extensive to normal form is presupposed (Levi, 1987). It would be a mistake to conclude from Hammond's argument that conditionalization, independence and ordering are all underwritten by strict consequentialism. Only independence is.

The reduction of extensive to normal form entailed by Hammond's consequentialism is presupposed by a great many economists, statisticians, philosophers and decision theorists interested in rational choice.[18] Arguments defending choice consistency and the weak ordering of options, the independence postulate, temporal conditionalization and other facets of Bayesianism have been based on some version of the reduction of extensive to normal form. For quite some time, Hammond has been both an advocate of the reduction thesis and an insightful investigator into its ramifications. (See, for example, Hammond, 1976 and 1977.)

The philosophical novelty introduced in Hammond's 1988 paper on this theme is the claim that consequentialism as presupposed by a wide variety of decision theorists including Savage assumes the reduction of extensive to normal form. I have argued that this claim is incorrect. Consequentialism claims that only consequences matter. Of course, consequentialists need to qualify this claim by taking into account the probabilities of consequences ensuring the implementation of options and, quite crucially, what is to count as an available option. There remains, of course, the widespread scepticism about the merits of consequentialism both in ethical theory and as a condition of rational choice. I share that scepticism regarding strict consequentialism.[19] Even so, there are many contexts where the conditions of strict consequentialism seem to be satisfied fairly well. The burden of this paper has been to argue that when planning future choices in a strict consequentialist setting, the question of what the decision maker ought to recognize as an available option presents a stumbling block to those who, like Hammond, seek to reduce sequential decision problems to nonsequential decision problems and derive strong conclusions about rational choice in strict consequentialist settings. The question of recognized availability has suffered uncritical neglect in studies of modern decision theory. The neglect is undeserved.

REFERENCES

Allais, M. (1953), "Le comportement de l'homme rationel devant le risque: Critique des postulats et axiomes de l'école americaine." *Econometrica* 21, pp. 503–46.

Balch, M. (1974), "On Recent Developments in Subjective Expected Utility Theory," in *Essays on Economic Behavior Under Uncertainty*, ed. by M. Balch, D. McFadden and S. Wu, North-Holland, pp. 45–54.

18 An important exception to this consensus among economists is M. Yaari (1977).

19 Weak consequentialism ought, however, to be noncontroversial and within the framework of weak consequentialism it remains possible to be a strict Bayesian (or, as I would prefer, to be a "quasi Bayesian" who abandons the requirements of ordering as well as the requirements of temporal conditionalization).

Fishburn, P. (1970). *Utility for Decision Making,* Wiley.

(1981), "Subjective Expected Utility: A Review of Normative Theories," *Theory and Decision* 13, pp. 139-99.

Gärdenfors, P. and Sahlin, N.-E. (1988), *Decision, Probability and Utility,* Cambridge U. Press.

Gibbard, A. and Harper, W. (1978), "Counterfactuals and Two Kinds of Expected Value," in *Foundations and Applications of Decision Theory,* ed. by C. Hooker, N. Leach and W. Harper, pp. 125-62, reprinted in *Decision, Probability and Utility,* ed. by P. Gärdenfors and N.-E. Sahlin (1988), Cambridge: Cambridge University Press.

Hammond, P. J. (1976), "Changing Tastes and Coherent Dynamic Choice," *Review of Economic Studies* 43, 159-73.

(1977), "Dynamic Restrictions on Metastatic Choice," *Economica* 44, 337-50.

(1988), "Consequentialist Foundations for Expected Utility," *Theory and Decision* 25, 25-78.

Jeffrey, R. C. (1965), *The Logic of Decision,* McGraw-Hill. 2nd rev. ed., U. of Chicago Press, 1983.

(1977), "A Note on the Kinematics of Preference," *Erkenntnis* 11, 135-41.

Kadane, J. B. and Larkey, P. D. (1982), "Subjective Probability and the Theory of Games," *Management Science* 28, 113-20.

Levi, I. (1974), "On Indeterminate Probabilities," *Journal of Philosophy* 71, 391-418, reprinted in *Decision, Probability and Utility,* ed. by P. Gärdenfors and N.-E. Sahlin (1988), Cambridge: Cambridge University Press.

(1977), "Subjunctives, Dispositions and Chances," *Synthese* 34, 423-55, reprinted in Levi (1984).

(1979), "Serious Possibility," *Essays in Honour of Jaakko Hintinkka,* Reidel, pp. 219-36, reprinted in Levi (1984).

(1980), *The Enterprise of Knowledge,* MIT Press.

(1984), *Decisions and Revisions,* Cambridge U. Press.

(1986a), *Hard Choices,* Cambridge U. Press.

(1986b), "The Paradoxes of Allais and Ellsberg," *Economics and Philosophy* 2, 23-54.

(1987), "The Demons of Decision," *The Monist* 70, 193-211.

(1988), "Iteration of Conditionals and the Ramsey Test," *Synthese* 76, 49-81.

(1996), *For the Sake of the Argument,* Cambridge: Cambridge U. Press.

Luce, R. D. and Krantz, D. (1971), "Conditional Expected Utility Theory," *Econometrica* 39, 253-71.

Luce, R. D. and Raiffa, H. (1958), *Games and Decisions,* Wiley.

MacCrimmon, K. R. and Larsson, S. I. (1979), "Utility Theory: Axioms vs. 'Paradoxes' " in *Expected Utility Hypotheses and the Allais Paradox,* ed. by M. Allais and O. Hagen, Reidel, pp. 333-409.

Machina, M. (1982), "Expected Utility Analysis without the Independence Axiom," *Econometrica* 50, 277-323.

Pettit, P. and Sugden, R. (1989), "The Backward Induction Paradox," *The Journal of Philosophy* 86, 169-82.

Savage, L. J. (1954), *The Foundations of Statistics,* Wiley.

Schick, F. (1979), "Self-Knowledge, Uncertainty and Choice," *British Journal for the Philosophy of Science* 30, 235-52, reprinted in Gärdenfors and Sahlin (1988).

Seidenfeld, T. (1988), "Decision Theory without 'Independence' or without 'Ordering': What Is the Difference?" *Economics and Philosophy* 4, 267-90.

Shackle, G. L. S. (1969), *Decision Order and Time,* Cambridge U. Press.

Spohn, W. (1977), "Where Luce and Krantz Do Really Generalize Savage's Decision Model," *Erkenntnis* 11, 113-34.

(1978), *Grundlagen der Entscheidungstheorie,* Scriptor Verlag.

Teller, P. (1973), "Conditionalization and Observation," *Synthese* 26, 218-38.

Yaari, M. (1977), "Endogenous Changes in Tastes: A Philosophical Discussion," *Erkenntnis* 11, 157-96.

5
Prediction, deliberation, and correlated equilibrium

In a pair of controversy provoking papers, Kadane and Larkey (1982, 1983) argued that the normative or prescriptive understanding of expected utility theory recommended that participants in a game maximize expected utility given their assessments of the probabilities of the moves that other players would make. They observed that no prescription, norm or standard of Bayesian rationality recommends how they should come to make probability judgments about the choices of other players. For any given player, it is an empirical question as to whether other players are Bayes rational, judge him to be Bayes rational, etc. just as it is an empirical question as to what the other players' goals and beliefs are. Participants in the game should, of course, use all empirical information available to them about how other players have behaved and how they are likely to behave in the given context of choice. Kadane and Larkey pointed out that rational players could participate in a game meeting these requirements without the game terminating with a Nash equilibrium or a result satisfying any other particular solution concept.

Kadane and Larkey did not deny that one could study how a participant in a game should make decisions if he made special assumptions about the other players - such as that they are expected utility maximizers. They emphasized, however, that no principle of rationality mandates that one rational player should believe that his opponents are rational. Consequently, unless it is true as a matter of empirical fact that participants in games who are expected maximizers have opinions that lead them to make choices satisfying the conditions of Nash equilibrium or some other such solution concept, the relevance of such solution concepts to either the making of recommendations to such participants or the descriptive study of their behavior is less obvious than it has seemed to many to be.

Consider, for example, a garden variety prisoner's dilemma as in Figure 5.1.

If Row is an expected utility maximizer, Kadane and Larkey recommend that he should make as careful an estimate as he can given his information of what Column is likely to do conditional on what he, Row,

2, 2	0, 6
6, 0	1, 1

Figure 5.1

chooses, compute the expectations of his options and choose the option maximizing expected utility. Row is supposed to be in a position to do this prior to actually reaching a decision when the determination of which option maximizes expected utility according to Row's values and beliefs can be of use to Row in reducing the set of options available to him (there are two) to a set of admissible options.

Consider the following two versions of what Row and Column might think about each other prior to deciding what to do.

Version 1: Row judges that what Column will do is probabilistically independent of what Row will do. In that case, it does not matter what specific probabilities Row assigns to Column's choices. The expected reward conditional on choosing Bottom is greater than the expected reward conditional on choosing Top, whatever Row's probability judgments about Top's choices might be. If Column makes similar judgments about what Row will do, Column should choose right and the payoffs are 1,1. Row and Column have achieved a Nash equilibrium.

However, a coherent expected utility maximizer might make different estimates of what Column will do.

Version 2: Row might judge that the probability that Column will choose left conditional on Row's choosing top and the conditional probability that Row will choose right conditional on Column's choosing bottom are both near 1. In that case, Row *should* choose Top. The expected utility conditional on choosing Top prior to making a choice is near 2, whereas the expected utility conditional on choosing Bottom is near 1. If Column's coherent estimates of what Row will do are probabilistically dependent on Row's choice in a parallel fashion, Column *should* choose left and the payoffs will be 2,2. Column and Row can both think about each other in this way and judge each other to be expected utility maximizers. Given that they do, they ought, according to expected utility theory, to make choices that fail to yield a Nash equilibrium.

In both versions of the prisoner's dilemma, Row and Column have rational beliefs and goals and their decisions are Bayes rational. Sometimes Bayes rational decisions lead to equilibrium and sometimes they don't.

Aumann's 1987 paper on correlated equilibrium is an explicit response to the scepticism registered by Kadane and Larkey regarding the use of solution concepts in game theory. Aumann declares his aim to be to show that "the notion of equilibrium is an unavoidable consequence" of the Bayesian view of the world (p. 2). He claims to show that Bayes rational players choose options in a way that realizes a "correlated equilibrium." Given certain assumptions about the attitudes of the players that Aumann holds to be reasonable, Column and Row must satisfy the conditions for a correlated equilibrium (that is, a certain kind of generalization of the notion of a Nash equilibrium). In particular, the following assumptions are made:[1]

A. *Joint Action Space Prior Assumption:* Prior to deciding which of the options available to him to exercise, each player i has a subjective or credal probability distribution π_i over hypotheses fully describing "states of the world including specification of the actions he and his opponent will make. In both versions of the prisoner's dilemma, there are four states in the action space and prior to reaching a decision, Column (Row) has a probability distribution π_c (π_r) over these four states.
B. *Common Knowledge of Rationality:* Prior to making their decisions, it is common knowledge among the players that each player will choose in a Bayes rational manner.
C. *Common Prior Assumption:* $\pi_c = \pi_r$
D. *Ratifiability Assumption:* (to be explained later).

If assumptions (A)-(D) are endorsed, the choices of Column and Row must satisfy the conditions for a correlated equilibrium. I have no quarrel with the elegant argument Aumann offers in favor of his claim. Aumann's conclusions do, indeed, follow from his premises. But I do not believe that his conclusions refute the Kadane-Larkey thesis because each one of these assumptions is untenable.

I shall first explain the content of what I am labeling the Ratifiability Assumption. My task will then be to argue that either the Ratifiability Assumption offers advice incompatible with the Bayesian dictum that rational agents choose options from those available to them that maximize expected utility relative to their probability and utility judgments or the roster of options available to a decision maker must satisfy some very restrictive and implausible conditions. According to the first alterna-

1 Throughout this discussion I will consider games between two players: Column and Row, or c and r. However, the issues raised generalize easily enough to games between n players.

tive, players who reach correlated equilibrium are not Bayes rational. According to the second, they may be Bayes rational under certain restricted circumstances but there will be other circumstances where Bayes rational players will not end up in a correlated equilibrium. These conclusions obtain whether or not assumptions (A)–(C) hold. Aumann's endorsement of ratifiability alone prevents him from responding relevantly to views such as those of Kadane and Larkey.

The discussion of ratifiability raises the issue of conditions under which options are available to players in a game. I shall argue that both assumptions (A) (and its corollary (C)) and (B) conflict with conditions on judgments of the availability of options. (A) and (B) both fail because prior to reaching a decision, a rational agent should not predict what he will do even though he may predict what others will do.

According to the Bayesian injunction to maximize expected utility, player i should choose as if he had considered every option σ available to him *prior* to reaching a decision and evaluated the expected value or payoff $E(\sigma)$ to i *using all the information available to him at that time.* This expected value is computed by multiplying the value of the outcome of i's choosing σ while the $n-1$ other players make a combination of moves $\mathbf{s}$ for every possible combination, by the probability $\pi(\mathbf{s}/\sigma)$ of that combination of moves occurring conditional on i's choosing σ. σ is admissible according to the injunction to maximize expected utility if and only if $E(\sigma) \geq E(\tau)$ for every available option τ.

This version of the injunction to maximize expected utility is an application of Fishburn's equation (3.5), which computes expected relative value for his "model 1" or "basic decision model" (Fishburn, 1964, p. 43). As Spohn (1977) notes, Fishburn's model 1 provides the most general formulation of the injunction to maximize expected utility advocated by Bayesians. To be sure, the task of furnishing a method of deriving the requisite probabilities and utilities from preferences among options is not fulfilled by specifying equation (3.5).[2] But as Spohn correctly observes, the question of the adequacy of expected utility theory as either a prescriptive or a descriptive theory does not stand or fall with the availability of an all purpose theory of elicitation from preferences among acts.

Having identified an admissible option, an agent i in control of himself will implement the recommended σ and regard the matter as settled having become convinced that σ is or will be implemented. In Aumann's

2 For efforts to provide the desired theories of elicitation, see Balch (1974); Balch and Fishburn (1974); Fishburn (1974); Luce and Krantz (1971); and the comments by Spohn (1977).

language, he discovers in which element of his "information partition" the actual state of the world is located. Updating his prior probability distribution over the joint action space to obtain a posterior distribution over the $n-1$ tuples of hypotheses concerning the combinations $\mathbf{s}$ of acts adopted by the other players given the option he has chosen, i can compute the expected value of his situation after having made his decision. Call this the *Bayes posterior expectation* $E_\sigma(\sigma)$ for the option σ once i has chosen σ and knows it. This expectation is computed using the probability distribution $\pi_\sigma(\mathbf{s})$ over hypotheses concerning what other players will choose relative to the information available to i after he has chosen σ and become certain that this move will be implemented. Conditionalization on the information that σ is implemented implies that $\pi_\sigma(\mathbf{s}) = \pi(\mathbf{s}/\sigma)$. On the assumption that i formed this expectation by conditionalizing on the information that σ is chosen and implemented, the Bayes posterior expectation $E_\sigma(\sigma)$ for σ once σ has been chosen should be equal to the Bayes expectation $E(\sigma)$ for σ as estimated prior to making a decision and coming to know which option is chosen.

Also, i can compute the expectation $E_\sigma(\tau)$ for another available option τ using for the computation $\pi_\sigma(\mathbf{s})$ and the payoffs for choosing τ instead of those for using σ. This expectation does not represent the expected value $E_\tau(\tau) = E(\tau)$ of τ when i has chosen τ and knows it unless $\pi_\tau(\mathbf{s}) = \pi(\mathbf{s}/\tau) = \pi(\mathbf{s}/\sigma) = \pi_\sigma(\mathbf{s})$. If this condition is satisfied for all available options, then hypotheses concerning what other agents will do are probabilistically independent of agent i's choices according to i's prior credal probability judgments (as in version 1 of the prisoner's dilemma). In that event, the following two conditions are equivalent:

Maximization of conditional expected utility: σ is admissible among the available options if and only if $E(\sigma) \geq E(\tau)$ for every available option τ.

Ratifiability: σ is admissible among the available options if and only if $E_\sigma(\sigma) \geq E_\sigma(\tau)$ for every available option τ.

Aumann stipulates that admissible options should satisfy the Ratifiability condition. This stipulation that options satisfying the Ratifiability condition are Bayes rational is the Ratifiability Assumption (D) announced above. The term "ratifiable" comes from Jeffrey (1983). Unlike Jeffrey, however, Aumann claims that Bayes rational choices are ratifiable choices.[3]

3 For Jeffrey, ratifiability is a supplement to the default conditional expected utility theory he favors. It applies when hands tremble – i.e., the agent is sure one option is chosen but is unsure as to which will be implemented. Aumann considers cases where the agent has chosen and is certain that his choice is (or will be) implemented. So the

As just noted, this claim is noncontroversial in contexts where states are probabilistically independent of acts as is required in the proposals made by Savage (1954) and Anscombe and Aumann (1963). Such independence does not hold in the second version of the prisoner's dilemma and in many other predicaments including cases that Aumann recognizes as realizing correlated equilibrium. Yet, Aumann seems to think that satisfying ratifiability is a necessary (and perhaps sufficient) condition for Bayes rational choice even when probabilistic independence fails. How can this requirement be squared with the demand that conditional expected value should be maximized in such cases? Maximizing conditional expected utility is a core requirement of the mainstream Bayesian tradition as the views of Luce and Krantz, Fishburn, and Spohn suggest.

A response is available to Aumann if cases of probabilistic dependence in games are treated as analogous to cases of moral hazard as understood by authors such as Drèze (1958, 1987). As Drèze understands moral hazard, the decision maker faces a choice between strategy-act pairs where an act from set A is representable as a function from states to (roulette lotteries) over prizes and a strategy is an action that modifies the probability distribution over the states by selecting one distribution from a given set Π. The options available to the decision maker are all (consistent) strategy-act pairs from $\Pi \times A$. If the purchaser of fire insur-

conditional expectations of the options not chosen are, for Aumann, "counterfactual" or, more accurately, "belief contravening" (see Aumann, 1987, p. 7). In spite of these differences, Aumann's criterion of Bayes rationality is formally similar to Jeffrey's notion of ratifiability in the respects spelled out in the text. Shin (1991, 1992) restricts the term "ratifiability" to trembling hand cases. He (1991) identifies a species of (modestly) ratifiable distributions with correlated distributions. He (1992) discusses the use of closest world counterfactual reasoning in order to obtain correlated equilibria. He is noncommittal as to whether the "trembling hand" or counterfactual interpretation better captures Aumann's view even though Aumann explicitly endorses the latter. He is tempted (1992, p. 413) to equate his reconstruction of Aumann in terms of trembling hand ratifiability with his counterfactual closest world reconstruction except for the fact that the closest world counterfactual account does not sit well within the Bayesian framework - as Jeffrey rightly points out. I argue later on that a Bayesian should not assign unconditional probabilities to action statements (as Jeffrey and Shin do) in contexts where choosing to implement one of them is optional for him. Even so, Shin is right to recognize that closest world accounts of conditionals are incompatible with Bayesianism. As explained later in the text, there is an account of ratifiable decision making that appeals neither to trembles nor to nearest world counterfactual reasoning and is compatible with maximizing conditional expected utility. So is the account of counterfactual reasoning I favor. See footnote 7. It is possible to interpret Aumann as charity requires - namely, as an orthodox conditional expected utility maximizer. This shall be explained shortly. I thank B. Skyrms for directing my attention to Shin's discussion of Aumann's version of ratifiability. Space does not permit a review of the literature on ratifiability - especially the extensive discussion in Skyrms (1990).

ance has the opportunity of insuring his property for more than its market value, he may consider whether or not also to take steps to increase the chance of a fire taking place. The option of purchasing the given policy decomposes into components: (a) purchasing the policy (an act) and (b) adopting a policy related to fire prevention or promotion controlling the chances of fire (a strategy). Options that maximize conditional expected utility among these strategy-act pairs can be determined as follows: (1) For each a in A, let $\Pi(a)$ be the set of strategy-act pairs $\langle\pi,a\rangle$ such that π maximizes the expected utility for a among all strategies in Π. (2) Choose any $[\pi,a]$ in a $\Pi(a)$ for which the expected utility is a maximum for all members of A.

In all such cases an optimal act relative to a strategy is going to meet the formal conditions for ratifiability. Maximizing conditional expected utility is also determinable as follows: (1') For each π in Π, let $A(\pi)$ be the set of acts that maximize expected utility relative to π among elements of A. (2') Choose a pair $[\pi,a]$ belonging to an $A(\pi)$ that maximizes expected utility among strategies or distributions in Π. In order that the Ratifiability Assumption be consistent with the injunction that expected utility be maximized, each element of A should be available relative to each distribution in Π.

Whereas Drèze envisages the two components of the strategy-act pair as chosen independently of one another, Aumann thinks of the set Π of strategies as a set of probability distributions over hypotheses as to what the other agents will do where each distribution π_i is conditioned on information e_i from a set E of possible outcomes of an experiment to be made available to the decision maker at the moment of decision. Once the agent finds out which e_i in E is true, he chooses among the available options. In so doing he is choosing a strategy-act pair. The decision maker has no choice of an element of Π, so step (2') is inoperative. But step (1') remains applicable.[4]

The scenario envisaged in the second version of the prisoner's dilemma is a predicament where hypotheses about what others will do (the states) are probabilistically dependent on acts. Nonetheless, the predicament cannot be understood coherently as a case of moral hazard in the sense of Drèze. Nor can maximizing conditional expected utility be made to fit the requirements of ratifiability as Aumann understands it.

4 Notice that on the assumption that the decision maker is Bayes rational, the conditional probability concerning what other players will do given the act chosen by the agent is equal to the conditional probability concerning what the other players will do given the information e_i.

Suppose that ρ_1 is Row's probabilistic prediction as to what Column will do conditional on Row's choosing top and ρ_2 is Row's probabilistic prediction conditional on Row's choosing bottom. ρ_1 assigns 0.9 to Column choosing left and 0.1 to Column choosing right, and ρ_2 assigns 0.1 to Column choosing left and 0.9 to Column choosing right. Let the set of strategies Π be the convex hull of these two probability distributions or just these two distributions. If version 2 of the prisoner's dilemma were analyzable as a case of moral hazard after the fashion of Drèze, [ρ_i, top] and [ρ_i, bottom] for $i = 1.2$ are options available to Row. Choosing bottom would be optimal against both distributions and, hence, is optimal among all available options. But maximizing conditional expected utility recommends that Row choose top in version 2 of the prisoner's dilemma. That is because Column simply does not have as options available to him [ρ_1, bottom] and [ρ_2, top] in version 2 but only [ρ_1, top] and [ρ_2, bottom]. Of these two, [ρ_1, top] maximizes conditional expected utility. That is to say, the set of options available in version 2 does not satisfy the requirements on the availability of options suggested by Drèze for moral hazard or those suggested by Aumann's version of the Ratifiability Assumption.

Thus, understanding ratifiability as consistent with maximizing conditional expected utility is purchased at the expense of ruling out as incoherent decision problems where decision makers face options as specified in version 2. But where is the argument for this?[5]

For my part, I think version 2 is coherent. So is version 1. We need adequate analyses of both of them. For Bayesians, the Drèze and Aumann approaches fail to account for version 2. This consideration alone suffices to call into question the Ratifiability Assumption (D) and with it Aumann's claim to have shown that the play of Bayes rational agents should reveal commitment to correlated equilibrium distributions as priors.

Perhaps, however, a more compelling and Bayesian argument may be invoked to rule out version 2. Row knows that Column will choose rationally by the Common Knowledge of Rationality Assumption (B). He also is convinced that Column judges that Row will choose top condi-

5 It does not matter whether version 2 predicaments are called "prisoner's dilemmas" or not. As I understand them, they are noncooperative single shot games where the players have not made any binding commitments beforehand. The players decide "independently" of one another in the sense that what one of them does is not controlled or influenced by what the other does. The players may believe all this coherently, may believe that their partners are Bayes rational and yet judge the decisions of their partners to be probabilistically dependent on their own decisions. Binmore (1992, 7.5.4) and other game theorists who quite sensibly keep so called "causal decision theory" at arm's length disagree for reasons I do not understand.

tional on Column choosing left with high probability and likewise with bottom conditional on right, because Row believes that Column shares a common prior with him and that Column's conditional probabilities are as specified in version 2. On this basis, Row should be certain that Column will choose left. But none of the probability distributions in the set Π for Row assigns probability 1 to Column choosing left. Here Row's convictions about Column's predictions concerning Row's choices together with Row's assumption that Column will choose rationally rule out version 2. Ratifiability need not be assumed as a condition of Bayes rational decision making. Version 2 can be ruled out by the Common Knowledge of Rationality Assumption. Perhaps, Aumann's invoking ratifiability was unnecessary in the first place.

Not so! Consider the following version of the prisoner's dilemma.

Version 3: Suppose Row judges that Column will play left given that Row plays top with probability 0.75 and will play left given that Row plays bottom with probability 0.1 and Column judges that Row will play top given that Column plays left with probability 0.75 and will play top given that Column plays bottom with probability 0.1 and Row and Column know this of each other. Then Row will judge that rational Column regards both choosing left and choosing right as maximizing expected utility and Column will pass a corresponding judgment regarding rational Row. If maximizing expected utility is all that is required for Bayes rationality, Row might well assign equal subjective probability to Column choosing left and choosing right. But the resulting joint distribution is not a correlated equilibrium distribution precisely because choosing top, though optimal, is not ratifiable.

In version 3, where more than one of Column's options is Bayes rational, Row's certainty that Column will choose rationally does not suffice to warrant certainty that Column will choose ratifiably. Common knowledge of rationality and a common prior without ratifiability does not yield a correlated equilibrium distribution prior. If we are to persuade Bayesians of the importance of equilibrium solution concepts for games, we face the daunting (in my judgment, insuperable) task of convincing them of the obligatory character of ratifiability as a condition of rational choice.

Aumann proves the general claim that if in an n-person game there is common knowledge of Bayes rationality with ratifiability built in (so that assumptions B and D obtain) and the Common Prior Assumption C is operative (so that the Joint Action Prior Assumption A holds as well), the common prior must yield a correlated equilibrium. This is the equilibrium concept that Aumann insists is inescapable as an ingredient in Bayesian views of rationality. He offers it as a response to the Kadane-

Larkey skepticism regarding the importance of equilibrium concepts in game theory from a Bayesian point of view.[6]

Although Aumann's formulation of standards of Bayes rationality (i.e., ratifiability) is retrospective so that it applies to the evaluation of options by agents once the choice has been made, his very argument for correlated equilibrium requires that options be assessable prospectively as well. Prospective prescriptions concerning what the agent ought to choose are made given the information *available to the decision maker* prior to actually choosing and implementing a course of action. We assume that the agent has sufficient computational capacity or logical omniscience and self knowledge of his beliefs and goals to be able to compute expected utilities of his options in the sense of expected utility relevant to identifying policy recommendations in the state prior to choice. The agent is supposed to then compare his evaluations with respect to expected utility for all the options the agent committed to judging available to him.[7]

Observe, however, that if an agent is certain prior to choice that he will not implement a certain policy, then from that agent's point of view at that time the policy is not optional even if we outside observers think the agent should not have been so certain. What we think the agent ought to believe is not relevant to our assessing the rationality of his judgments as long as his beliefs meet the standards of weak rationality

6 A correlated equilibrium distribution over the joint action space of all the players is one where for every player i and every option σ available to i according to the game assigned positive prior probability and every τ available to i according to the game, $E_{\sigma(\sigma)} \geq E_{\sigma}(\tau)$. Formal definitions are found in Aumann (1974) and (1987) together with discussion of intended interpretations.

7 Shin (1992) rightly observes that if an agent has already chosen an option, any comparison of the expected value of the option chosen with other options will have to be "counterfactual" or, more accurately, "belief contravening." The agent needs to suppose that instead of choosing the option he chose, he had chosen one of the other options. According to Bayesian approaches to decision making, the available options are the options available *according to the agent's perspective prior to making a decision.* So the suppositional reasoning of the agent invites him first to contract his state of full belief to the state he was in prior to reaching a decision and then expand by adding an option judged to be available relative to that *prospective* belief state. This kind of suppositional reasoning differs from the "closest world" reasoning Shin favors. It recommends maximizing act conditional expected utility from the point of view prior to choice just as Bayesian decision makers are urged to do. Whatever the merits of Shin's account of counterfactual reasoning in the context of game theory might be, it is *not* congenial with Bayesian thinking. If Aumann understands maximizing expected utility in the manner sketched by Shin, his paper is not a response to the views of Kadane and Larkey or other Bayesian decision theorists skeptical of the centrality of the concept of equilibrium. I have sought to interpret Aumann's use of ratifiability in a manner compatible with what appear to be his purposes. For elaboration of an account of conditionals consonant with Bayesian decision theory see Levi (1996).

that theories of expected utility require. If an agent is certain that he will not perform a certain course of action, the agent is obliged to rule out choosing that course of action as available for him as an option. It is not irrational for him, therefore, to fail to choose that course of action even if he judges that if it were an available option for him, it would have been better for him to choose it than what, given what he believes, is best.

Recall that retrospectively options need to be ratifiable to be rationally chosen according to Aumann. Prospectively, there must have been a prior that is a correlated equilibrium distribution. In the case of the prisoner's dilemma, only version 1 is allowed and in that case, both Row and Column must be certain as to which option they will pick prospectively – i.e., prior to making a decision. But in that case, neither Row nor Column has a decision to make. Each has but one available option. Row does not have Top available to him as an option and Column does not have Left. At best we can say that deliberation aimed at identifying an admissible subset of options from among the available ones given the decision maker's goals and beliefs is vacuous. It recommends what is going to happen anyhow. In that case, there is no need for Aumann to invoke ratifiability in order to compare a given option with alternatives. There is only one option available.

There are gamelike situations where maximizing conditional expected utility ratifiably allows more than one option to be consistent with the decision maker's information prior to reaching a decision. (Version 3 of the prisoner's dilemma is an example.) However, the Assumption of Common Knowledge of Bayes Rationality (Assumption (B)) implies that each decision maker is certain that he will choose rationally so that from the decision maker's point of view every option available to him is Bayes rational. In this setting, the player cannot appeal to Bayes rationality to reduce the set of options available to him to an admissible proper subset for choice. Such reduction is always vacuous. Game theory is often advertised as taking into account how agents do or should reason to a decision taking into account the reasoning of others. Assumption (B) precludes such reasoning from leading to anything but the vacuous recommendation: Choose any one of the options available to you!

If Row and Column are to use the information available to them prior to choice to deliberate as to which of their respective options to choose, they certainly ought to have some idea of what their standards for evaluating options are and should be able to identify which of their options are the Bayes rational choices. But they need not assume that they will choose rationally and implement the Bayes rational recommendations. Indeed, if they do not do so, then they are not to that extent prevented from recognizing acts other than Bayes rational acts as available to them. But if a player does not take for granted that he will act rationally, he cannot have common knowledge that all players will act rationally. An outside observer can assume that all players will act rationally but, if deliberation is not to be vacuous, cannot attribute to the players conviction that they personally will act rationally. Thus, in addition to doubts raised about the Ratifiability Assumption (D), the assumption of common knowledge of Bayes rationality (B) has been called into question. And it is called into question without directly objecting to the Ratifiability condition.

Assumption (A) - The Joint Action Space Prior Assumption - implies that each player assigns unconditional credal probabilities to each of his options prior to choice. However, this too leads to trouble, as W. Spohn (1977, 1978) has already observed. As I have argued (Levi, 1991), allowing unconditional probabilities to be assigned to options as Assumption (A) requires leads to the same embarrassing vacuous recommendation as endorsing Assumption (B) does. The rational agent must be certain prior to choice that he will choose rationally. The principles of rational choice cannot be used by the deliberating agent as standards for criticizing and evaluating his options before deciding. To avoid returning to Assumption (B), we must reject Assumption (A) as well.

Since Assumption (C) - the Common Prior Assumption - presupposes Assumption (A), it too is called into question.

The objections raised against (A)-(C) apply whether one insists on the Ratifiability Assumption (D) at the outset or not. No decision maker should be certain that he will choose rationally. No decision maker should assign unconditional probabilities to hypotheses as to what he will choose. Otherwise deliberation is vacuous. It is not enough to reject correlated equilibrium distributions as priors over joint action spaces. One should abandon the idea of priors over such joint spaces altogether.

Aumann himself addresses the issue of treating "personal choice as a state variable" (Aumann, 1987, pp. 8-9). The difficulty he addresses is that treating personal choice as a state variable deprives each participant in the game of freedom of choice. He rightly sees that a game theorist or "outside observer" could regard the participants' decisions as state variables without depriving any participant of "freedom of choice" *precisely because the outside observer is not a participant in the game and is not in the business of predicting his own choices.*

This correct observation does not, however, address the point under discussion here. If the outside observer attributes minimal rationality to the players of the game, he should not model them as endorsing unconditional probability distributions over their own acts. He can make predictions as to how each of them will choose; but no participant in the game should make predictions as to how he will choose or whether he will choose rationally and the outside observer, who assumes that each agent is rational, should not attribute such predictions to participants. But if the outside observer is going to derive a correlated equilibrium from his modeling assumptions and Aumann's correlated equilibrium theorem, the outside observer must assume that each of the participants has an unconditional probability distribution prior to choice over the joint space of acts for the players. My contention is that if the outside observer does this, he is committed to denying that the players are capable of using the principles of rational choice to reduce the set of available options to a proper subset of admissible ones.

The issue is not whether an outside observer can regard the participants' decisions as state variables while taking these agents to be free to choose. According to those variants of so-called soft determinist doctrine with which I am sympathetic, an outside observer can do this. But my weakened version of Spohn's thesis does not contradict such soft determinism. I question whether agent X can regard himself as free to choose in the sense relevant to his deliberation and regard his choice as a state variable. If X cannot coherently do so, the outside observer should not model X or any other player in the game as doing so.[8]

Aumann agrees with Kadane and Larkey that models of games should cohere with expected utility theory. He complains, however, that Kadane and Larkey "ignored the fact that a rational player must take into account how other players reason about him." In my opinion, the shoe is on the

8 Aumann and Brandenburger (1994) examine conditions under which the "conjectures" of players in a game as to what other players will do constitute a Nash equilibrium. Whatever the significance of Nash equilibria in conjectures might be, it does not seem to bear on the issues raised by Kadane and Larkey. But it is worth comparing the assumptions used for the main theorems of this paper with the assumptions used in the correlated equilibrium paper. Theorem A claims that mutual knowledge of the payoff functions, rationality of the players and conjectures are sufficient for Nash equilibrium of conjectures in the 2 person case. Mutual knowledge of rationality presupposes that each player knows that he is rational and, hence, that the difficulty raised in the text arises. In the n-person case, theorem B states that in addition to the conditions just mentioned, a common prior is available, and that the conjectures are commonly known. So the problems related to our Assumptions (A) and (C) resurface. The common prior raises another complication as well. The scheme examined by Aumann and Brandenburger stipulates that the states over which the common prior is defined includes specifications of each player's conjectures concerning what the other players will do. If an agent has a prior over such states, he must be uncertain when in that prior state as to what his conjecture is. But this means that *in that prior state* he is unsure of his probabilities. As Savage (1954) pointed out, this leads to incoherence.

other foot. As Kadane and Seidenfeld (1992) illustrate in subtle and interesting detail, player i can utilize whatever information i may have about the reasoning of other players concerning what i will do in making predictions as to how the other players will choose. If i is rational, however, not only should i avoid predicting his own choice; but in reasoning about how other rational agents reason about him, he should not suppose that they attribute to him predictions about his own choices. If Spohn's thesis is correct (as it certainly appears to be), it is Aumann and not Kadane and Larkey who ignores "the fact that a rational player must take into account how other players reason about him."

REFERENCES

Anscombe, F. J. and Aumann, R. J. (1963), "A Definition of Subjective Probability," *Annals of Mathematical Statistics* 35, 199-205.

Aumann, R. J. (1974), "Subjectivity and Correlation in Randomized Strategies," *Journal of Mathematical Economics* 1, 67-96.

(1987), "Correlated Equilibrium as an Expression of Bayesian Rationality," *Econometrica* 55, 1-18.

Aumann, R. J. and Brandenburger, A. (1994), "Epistemic Conditions for Nash Equilibrium," Discussion paper #94 of the Center for Rationality and Interactive Decision Theory of the Hebrew University of Jerusalem (to be published in *Econometrica*).

Balch, M. (1974), "On Recent Developments in Subjective Expected Utility Theory," in M. Balch, D. McFadden and S. Wu (1974), pp. 45-54.

Balch, M. and Fishburn, P. C. (1974), "Subjective Expected Utility for Conditional Primitives," in M. Balch, D. McFadden and S. Wu (1974), pp. 57-69.

Balch, M., McFadden, D. and Wu, S. (1974) (eds.), *Essays on Economic Behavior Under Uncertainty,* Amsterdam: North-Holland.

Binmore, K. (1992), *Fun and Games,* London: D.C. Heath.

Drèze, J. (1958), *Individual Decision Making Under Partially Controllable Uncertainty,* unpublished Ph.D. thesis, Columbia University.

(1987), *Essays on Economic Decisions Under Uncertainty,* Cambridge: Cambridge University Press.

Fishburn, P. C. (1964), *Decision and Value Theory,* New York: Wiley.

(1974), "On the Foundations of Decision Making Under Uncertainty," in M. Balch, D. McFadden and S. Wu (1974), pp. 25-44.

(1981), "Subjective Expected Utility: A Review of Normative Theories," *Theory and Decision* 13, 139-199.

Jeffrey, R. C. (1965), *The Logic of Decision,* New York: McGraw-Hill.

(1983), *The Logic of Decision,* 2nd ed., Chicago: University of Chicago Press.

Kadane, J. B. and Larkey, P. D. (1982), "Subjective Probability and the Theory of Games," *Management Science* 28, 113-20.

(1983), "The Confusion of Is and Ought in Game Theoretic Contexts," *Management Science* 29, 1365–79.

Kadane, J. B. and Seidenfeld, T. (1992), "Equilibrium, Common Knowledge and Optimal Sequential Decisions," *Knowledge, Belief and Strategic Interaction,* edited by C. Bicchieri and M. L. dalla Chiara, Cambridge: Cambridge U. Press, pp. 27–45.

Levi, I. (1991), "Consequentialism and Sequential Choice," *Foundations of Decision Theory,* edited by M. Bacharach and S. Hurley, pp. 92–122.

(1996), *For the Sake of the Argument,* Cambridge: Cambridge University Press.

Luce, R. D. and Krantz, D. H. (1971), "Conditional Expected Utility," *Econometrica* 39, 253–71.

Savage, L. J. (1954), *Foundations of Statistics,* New York: Wiley; 2nd revised edition, New York: Dover, 1972.

Shin, H. S. (1991), "Two Notions of Ratifiability," *Foundations of Decision Theory,* edited by M. Bacharach and S. Hurley, Oxford: Basil Blackwell, pp. 242–262.

(1992), "Counterfactuals and a Theory of Equilibrium in Games," *Knowledge, Belief and Strategic Interaction,* edited by C. Bichhieri and M. L. Dalla Chiara, Cambridge: Cambridge University Press, pp. 397–413.

Skyrms, B. (1990), "Ratifiability and the Logic of Decision," *Midwest Studies in Philosophy* 15, 44–55.

Spohn, W. (1977), "Where Luce and Krantz Do Really Generalize Savage's Decision Model," *Erkenntnis* 11, 113–34.

(1978), *Grundlagen der Entscheidungstheorie,* Scriptor Verlag.

6
On indeterminate probabilities

Some men disclaim certainty about anything. I am certain that they deceive themselves. Be that as it may, only the arrogant and foolish maintain that they are certain about everything. It is appropriate, therefore, to consider how judgments of uncertainty discriminate between hypotheses with respect to grades of uncertainty, probability, belief, or credence. Discriminations of this sort are relevant to the conduct of deliberations aimed at making choices between rival policies not only in the context of games of chance, but in moral, political, economic, or scientific decision making. If agent X wishes to promote some aim or system of values, he will (*ceteris paribus*) favor a policy that guarantees him against failure over a policy that does not. Where no guarantee is to be obtained, he will (or should) favor a policy that reduces the probability of failure to the greatest degree feasible. At any rate, this is so when X is engaged in deliberate decision making (as opposed to habitual or routine choice).

Two problems suggest themselves, for philosophical consideration:

Work on this essay was partially supported by N.S.F. grant GS 28992. Research was carried out while I was a Visiting Scholar at Leckhampton, Corpus Christi, Cambridge. I wish to thank the Fellows of Corpus Christi College and the Departments of Philosophy and History and Philosophy of Science, Cambridge University, for their kind hospitality. I am indebted to Howard Stein for his help in formulating and establishing some of the results reported here. Sidney Morgenbesser, Ernest Nagel, Teddy Seidenfeld, and Frederic Schick as well as Stein have made helpful suggestions.

This version of my 1974 paper includes the following revisions of the earlier paper: (i) A correction of the convexity condition; (ii) an improvement of the condition of confirmational conditionalization; (iii) an alteration of the substance of footnotes 3 and 4 to reflect more accurately the position I took in subsequent publications; and (iv) some relatively minor stylistic revisions of no substantive import.

In his interesting discussion of this paper, S. Spielman (1975) correctly points out that the view adopted in this paper represents a change from the position I took in 1970. In the earlier paper, I sought to reduce all revisions of probability judgment to revisions of knowledge in accordance with conditionalization. This paper marks a shift from that position.

From *The Journal of Philosophy* LXXI, 13 (July 18, 1974), pp. 391–418. Reprinted by permission of *The Journal of Philosophy.*

The problem of rational credence: Suppose that an ideally rational agent X is committed at time t to adopting as certain a given system of sentences $K_{X,t}$ (in a suitably regimented L) and to assigning to sentences in L that are not in $K_{X,t}$, various degrees of (personal) probability, belief, or credence. The problem is to specify conditions that X's "corpus of knowledge" $K_{X,t}$ and his "credal state" $B_{X,t}$ (i.e., his system of judgments of probability or credence) should satisfy in order to be reasonable.

The problem of rational choice: Given a corpus $K_{X,t}$ and a credal state $B_{X,t}$ at t, how should X make decisions between alternative policies from which he must choose one at t?

Consideration of these two problems should lead to examination of a third. A rational agent X is entitled to count as certain at t not only logical, mathematical, and set-theoretical truths supplemented by suitably produced testimony of the senses, but theories, laws, and statistical claims as well. At the same time, the revisability of X's corpus at t should be recognized not only by others but by X himself. Moreover, just as X's judgments of certainty are liable to revision, so too are his judgments of probability or credence. Indeed, the two types of modification are apparently interdependent, and this interdependence itself deserves examination. The third problem, therefore, is as follows:

The problem of revision: Under what conditions should X modify his corpus $K_{X,t}$ or his credal state $B_{X,t}$, and, if he should do so, how should he choose between alternative ways of making revisions?

In this essay, I shall not attempt to solve the problem of revision. However, I shall indicate how a *prima facie* obstacle to offering anything other than a dogmatic or antirationalistic answer to the question can be eliminated.

The obstacle is a serious one; for it derives from a very attractive system of answers to the problem of rational credence and the problem of rational choice. I allude to what is called the "bayesian" view. Bayesians do not agree with one another in their answers to these questions in all respects. The views of Harold Jeffreys and the early views of Rudolf Carnap are not consonant in important ways with the ideas of Bruno de Finetti and Leonard J. Savage (or the later Carnap). Nonetheless, the answers these and a host of other authors offer to the first two questions share certain important ramifications for the problem of revision. One of these implications is the commitment to either dogmatism or antirationalism.

Of course, identifying an objectionable consequence of bayesianism, where the objection is grounded on a question of philosophical principle, is in itself unlikely to persuade devoted bayesians to abandon their position. Such authors will be tempted to modify philosophical principle so as to disarm the objection; and they will have good reasons for doing so. Bayesian doctrine does offer answers to the first two questions. These answers are derivable from a system of principles which are precise and simple. Even the disputes between bayesians can be formulated with considerable precision. Furthermore, the prescriptions bayesians recommend for making choices appear to conform to presystematic judgment at least in some contexts of decision. Rival attempts to answer the problems of rational credence and rational choice seem either eclectic or patently inadequate when compared with the bayesian approach.

Thus, it is not enough to complain of the defects of bayesianism. The serious challenge is to construct an alternative system of answers to the problems of rational credence and choice which preserves the virtues of bayesianism without its vices - in particular, the defects it exhibits relevant to the problem of revision.

In this paper, I shall outline just such a rival view.

1. X's corpus of knowledge $K_{X,t}$ at t identifies a set of options A_1, A_2, . . . , A_n as the options from which he will choose (at t' identical with or later than t) at least and at most one. In addition, $K_{X,t}$ implies that at least and at most one of the hypotheses h_1, h_2, . . . , h_m is true and that each of the h_j's is consistent with $K_{X,t}$. Finally, $K_{X,t}$ implies that, if X chooses A_i when h_j is true, the hypothesis o_{ij} asserting the occurrence of some "possible consequence" of A_i is true.

The problem of rational choice is to specify criteria for evaluating various choices of A_i's from among those feasible for X according to what he knows at t. Such criteria may be construed as specifying conditions for "admissibility." Option A_i is admissible if and only if X is permitted as a rational agent to choose A_i from among the feasible options. If A_i is uniquely admissible, X is obliged, as a rational agent, to choose it. In general, however, unique admissibility cannot be guaranteed, and no theory of rational choice pretends to guarantee it.

Bayesians begin their answer to the problem of rational choice by assuming that X is an ideally rational agent in the following sense:

i. X has a system of evaluations for the possible consequences (the o_{ij}'s) representable by a real-valued "utility" function $u(o_{ij})$ unique up to a positive affine transformation (i.e., where utility assignments are nonarbitrary once a 0 point and a unit are chosen - as in the case of measuring temperature).

ii. X has a system of assignments of degrees of credence to the o_{ij}'s given the choice of A_i representable by a real-valued function $Q(o_{ij}; A_i)$ conforming to the requirements of the calculus of probabilities. Often X will assign credence values to the "states of nature" $h_1, h_2, \ldots, h_n$ so that the h_j's are probabilistically independent of the option chosen. When this is so, $Q(o_{ij}, A_i)$ equals the unconditional credence (given $K_{X,t}$) $Q(h_j)$. In the sequel, I shall suppose that we are dealing with situations of this kind.

Given such a utility function $u(o_{ij})$ and Q-function $Q(h_j)$, let

$$\sum_{j=1}^{m} u(o_{ij})Q(h_j).$$

$E(A_i) = E(A_i)$ is the expected utility of the option A_i.

Bayesians adopt as their fundamental principle of rational choice the principle that an option is admissible only if it bears maximum expected utility among all the feasible options.

Very few serious writers on the topic of rational choice object to the principle of maximizing expected utility in those cases where X's values and credal state can be represented by a utility function unique up to a positive affine transformation and a unique probability function. The doubts typically registered concern the applicability of this principle. That is to say, critics doubt that ordinary men have the ability under normal circumstances to satisfy the conditions of ideal rationality stipulated by strict bayesians even to a modest degree of approximation.

The bayesian riposte to doubts about applicability is to insist that rational men should meet the requirements for applying the principle of maximizing expected utility and that, appearances to the contrary notwithstanding, men are quite capable of meeting these requirements and often do so.

I am not concerned to speculate on our capacities for meeting strict bayesian requirements for credal (and value) rationality. But even if men have, at least to a good degree of approximation, the abilities bayesians attribute to them, there are many situations where, in my opinion, rational men *ought not* to have precise utility functions and precise probability judgments. This is to say, on some occasions, we should avoid satisfying the conditions for applying the principle of maximizing expected utility even if we have the ability to satisfy them.

In this essay, reference to the question of utility will be made from time to time. I shall not, however, attempt to explain why I think it is sometimes (indeed, often) irrational to evaluate consequences by means of a utility function unique up to a positive affine transformation. My chief concern is to argue that rational men should sometimes avoid adopting numerically precise probability judgments.

The bayesian answer to the problem of rational choice presupposes at least part of an answer to the problem of rational credence. For a strict bayesian, a rational agent has a credal state representable by a numerically precise function on sentences (or pairs of sentences when conditional probability is considered) obeying the dictates of the calculus of probabilities.

There are, to be sure, serious disputes among bayesians concerning credal rationality. In his early writings, Carnap (1962, pp. 219-241) believed that principles of "inductive logic" could be formulated so that, given *X*'s corpus $K_{X,t}$, *X*'s credal state at *t* would be required by the principles of inductive logic to be represented by a specific *Q*-function that would be the same for anyone having that corpus. Others (including the later Carnap, 1971a, p. 27) despair of identifying such strong principles. Nonetheless, bayesian critics of the early Carnap's program for inductive logic continue to insist that ideally rational agents should assign precise probabilities to hypotheses.

2. *X*'s corpus of knowledge $K_{X,t}$ shall be construed to be the set of sentences (in *L*) to whose certain truth *X* is committed at *t*. I am not suggesting that *X* is explicitly or consciously certain of the truth of every sentence in $K_{X,t}$, but only that he is committed to being certain. *X* might be certain at *t* of the truth of *h* and, hence, be committed to being certain of $h \vee g$, without actually being certain. Should it be brought to *X*'s attention, however, that $h \vee g$ is a deductive consequence of *h*, he would be obliged as a rational agent either to cease being certain of *h* or to take $h \vee g$ to be certain. The latter alternative amounts to retaining his commitment; the former to abandoning it.

In this sense, *X*'s corpus of knowledge at *t* should be a deductively closed set of sentences. Insofar as we restrict our attention to changes in knowledge and credence which are changes in commitments, modifications of corpora of knowledge are shifts from deductively closed sets of sentences to other deductively closed sets of sentences. Such modifications come in three varieties:

1. *Expansions,* where *X* strengthens his corpus by adding new items. Some examples of expansion are acquiring new items via observation, from the testimony of others and through inductive or nondeductive inference leading to the "acceptance" of statistical claims, laws, or theories into the corpus.

2. *Contractions,* where X weakens his corpus by removing items. This can happen when X detects an inconsistency in his corpus due to his having added at some previous expansion step an observation report that contradicts assumptions already in his corpus, or when X finds himself in disagreement with Y (whose views he respects on the point at issue) and wishes to resolve the dispute without begging the question.
3. *Replacements,* where X shifts from a theory containing one assumption to another containing an assumption contradicting the first. This can happen when X substitutes one theory for another in his corpus.

No matter which kind of modification is made, I shall suppose that there is a "weakest" potential corpus *UK* (the "urcorpus") of sentences in L such that no rational agent should contract that corpus. *UK* is the deductively closed set of sentences in L such that every potential corpus in L is an expansion of *UK* (or is *UK* itself). I shall suppose that *UK* contains logical truths, set-theoretical truths, mathematical truths, and whatever else might be granted immunity from removal from the status of knowledge. (The items in *UK* are in this sense incorrigible.)

Replacement poses special problems for an account of the revision of knowledge. At t when X's corpus is $K_{X,t}$, why should he shift to a corpus K^* which is obtained by deleting items from $K_{X,t}$ and replacing them with other items inconsistent with the first? From X's point of view, at t, he is replacing a theory which he is certain is true by another which he is certain is false.

The puzzle can be avoided by regarding replacements for purposes of analysis as involving two steps: (a) contraction to a corpus relative to which no question is begged concerning the rival theories, and (b) subsequent expansion based on the information available in the contracted corpus, supplemented, perhaps, by the results of experiments conducted in the interim.

Those who insist on attempting to justify replacements without decomposing them into contractions followed by expansions confront the predicament that they cannot justify such shifts without begging questions. Such justification is no justification. The conclusion that beckons is that all replacements are forms of "conversion" to which men are subjected under revolutionary stress. This is the view which Thomas Kuhn has made so popular and which stands opposed to views that look to the formulation of objective criteria for the evaluation of proposed modifications of knowledge.

3. How does all this relate to bayesian views about the revision of credal states?

Consider *X*'s corpus of knowledge $K_{X,t}$ at *t*. *X*'s credal state $B_{X,t}$ at *t* is, according to strict bayesians, determined by $K_{X,t}$. Strict bayesians disagree among themselves concerning the appropriate way in which to formulate this determination. The following characterization captures the orthodox view in all its essentials.

Let *K* be any potential corpus (i.e., let it be *UK* or an expansion thereof). Let $C_{X,t}$ (*K*) be *X*'s judgment at *t* as to what his credal state should be were he to adopt *K* as his corpus of knowledge. I shall suppose that *X* is committed to judgments of this sort for every feasible *K* in *L*. The resulting function from potential corpora of knowledge to potential credal states shall be called *X*'s "confirmational commitment" at *t*.

According to strict bayesians, no matter what corpus *K* is (provided it is consistent), $C_{X,t}$ (*K*) is representable by a probability function where all sentences in *K* receive probability 1. In particular, $C_{X,t}$ (*UK*) is representable by a function *P* (*x;y*) - which I shall call a *P*-function, to contrast it with a *Q*-function representing $C_{X,t}(K)$ where *K* is an expansion of *UK*.

Strict bayesians adopt the following principle, which imposes restrictions upon confirmational commitments:

Confirmational conditionalization: If *K* is obtained from *UK* by adding *e* (consistent with *UK*) to *UK* and forming the deductive closure, *P*(*x;y*) represents $C_{X,t}(UK)$, and *Q*(*x;y*) represents $C_{x,t}(K)$, $Q(h;f) = P(\text{h};f\&e)$.

In virtue of this principle, *X*'s confirmational commitment is defined by specifying $C_{X,t}(UK) = C_{X,t}$ and employing confirmational conditionalization.[1] *X*'s credal state at *t*, $B_{X,t}$, is then determined by $K_{X,t}$ and $C_{X,t}$ according to the following principle:

Total Knowledge: $C_{X,t}(K_{X,t}) = B_{X,t}$.

Notice that the principle of confirmational conditionalization, even when taken together with the principle of total knowledge, does not prescribe how *X* should modify his credal state given a change in his corpus of knowledge.

1 Confirmational commitments built on the principle of confirmational conditionalization are called "credibilities" by Carnap (1971a, pp. 17-19). The analogy is not quite perfect. According to Carnap, a credibility function represents a permanent disposition of *X* to modify his credal states in light of changes in his corpus of knowledge. When credibility is rational, it can be represented by a "confirmational function." Since I wish to allow for modifications of confirmational commitments as well as bodies of knowledge and credal states, I assign dates to confirmational commitments. Throughout I gloss over Carnap's distinction between credibility functions and confirmation functions (1971a, pp. 24-27).

To see this, suppose that at t_1 X's corpus is K_1 and that at t_2 his corpus K_2 is obtained from K_1 by adding e (consistent with K_1) and forming the deductive closure. From confirmational conditionalization and total knowledge, we can conclude that *if X does not alter his confirmational commitment in the interim from* t_1 *to* t_2, then, if Q_1 represents $B_{X,t}$, and Q_2 represents $B_{X,t}$, $Q(h;f) = Q_1(h;f\&e)$. Should X renege at t_2 on the confirmational commitment he adopted at t_1, the change in knowledge just described need not and will not, in general, lead to a modification of credal state of the sort indicated.

Nonetheless, strict bayesians unanimously suppose that a rational agent will, save under unusual circumstances, modify his credal state in the fashion indicated. This mode of revising credal states is often called "conditionalization." To distinguish it from confirmational conditionalization and other types of conditionalization, I shall call it "intertemporal credal conditionalization." I contend that the strict bayesian endorsement of intertemporal credal conditionalization presupposes commitment to the following principle:

Confirmational tenacity: For every X, t, and t', $C_{X,t} = C_{X,t'}$.

Thus, strict bayesians have an answer to the problem of revising credal states. X's confirmational commitment is to be held fixed over time. Given such a fixed commitment, the credal state he should adopt is determined for each possible modification of his corpus of knowledge which is a consistent expansion of *UK*. The problem of revising credal states reduces, therefore, to the problem of revising corpora of knowledge.

Is this answer to the problem of revision satisfactory? It would be, in my opinion, if the program for inductive logic envisaged by Carnap in his early writings on the subject could be realized. Inductive logic would then be strong enough to single out a standard *P*-function that all rational agents should adopt as their confirmational commitment. A fortiori, all such agents should hold that commitment fast at all times.

Few bayesians now think an inductive logic of the requisite power can be constructed. Their reasons (which, in my opinion, are sound) need not detain us. In response to this skepticism, most bayesians no longer require that all rational agents endorse a single standard confirmational commitment. They hold that rational X is perfectly free to pick any confirmational commitment consonant with the principles of inductive logic. Rational Y is quite free to pick a different commitment. However, bayesians tend to insist that, once X and Y have chosen their respective commitments, they should hold them fixed. To do this is to follow the probabilistic analogue of the method of tenacity so justly criticized by Peirce in "Fixation of Belief."

In the spirit of Peirce, it would have been far better to say that a rational X should not modify his confirmational commitment capriciously – i.e., without justification. To follow this approach, however, demands consideration of criteria for justified modifications of confirmational commitments. Bayesians not only fail to do this, but, as I shall now argue, they cannot do so without great difficulty. Given the bayesian answer to the problem of rational credence, no shift can be justified. If I am right, for bayesians, either tenacity should be obeyed, or, if not, justification is gratuitous. I think this implication of bayesian doctrine is to be deplored and should lead to scrutiny of other approaches.

4. Modifying a confirmational commitment is not quite the same as modifying a corpus of knowledge. Yet, shifting from a confirmational commitment represented by a precise probability function to another confirmational commitment represented by a different precise probability function seems analogous to replacement in the following sense: The shift from confirmational commitment C_1 to confirmational commitment C_2 involves a shift to a confirmational commitment conflicting with C_1 in the sense that the *P*-function X uses to determine his credal state relative to his corpus when C_1 is adopted yields different precise subjective probability or credence assignments for hypotheses from those which X would make were he to adopt C_2 (and keep his corpus constant).

From X's vantage point at t when he endorses C_1, C_2 is illegitimate. He cannot justify shifting to C_2. At least, he cannot justify a direct shift. Can he do so indirectly by first performing a shift analogous to contraction from C_1 to C_3, which begs no questions concerning the merits of C_1 and C_2? Not from a strict bayesian point of view; for C_3 would, like C_1 and C_2, have to be representable by a precise *P*-function. The shift from C_1 to C_3 would be as problematic as the shift from C_1 to C_2.

Thus, from a bayesian point of view, no shift from one confirmational commitment to another can be justified. A rational man should conform to confirmational tenacity so that no justification is needed or else hold that some shifts are permitted without justification. Carnap (1971b, 51–52) sometimes seems to recognize shifts in confirmational commitments as a result of conceptual change. Alternatively, one might allow shifts in confirmational commitment due to conversion under revolutionary stress. Except for the minimal requirement that the shift be to a commitment obeying requirements of inductive logic, no critical control is to be exercised. Bayesians are committed to being dogmatically tenacious or arbitrarily capricious.

The source of the difficulty should be apparent. Bayesians restrict the confirmational commitments a rational agent may adopt to those representable by numerically precise probability functions. This precludes shifting from a confirmational commitment C_1 to a confirmational commitment C_3 that begs no questions as to the merits of C_1 and another commitment C_2 that conflicts with C_1. My thesis is that not only are rational men allowed to make shifts to non-question-begging commitments but that on many occasions they ought to do so. That is to say, it is sometimes appropriate for a rational agent to adopt a confirmational commitment that is indeterminate in the sense that it cannot be represented by a numerically precise probability function. If we relax the stringent requirements imposed by bayesians on confirmational commitments and credal states so as to allow for such shifts, there is at least some hope that we can avoid endorsement of tenacity or capriciousness. Within the strict bayesian framework, we cannot expect to do so except by clinging desperately to Carnap's early program for constructing an inductive logic so strong as to single out a standard *P*-function to represent the uniquely rational confirmational commitment (for a given language).

I propose to explore one way of relaxing strict bayesian requirements. The basic idea is to represent a credal state (confirmational commitment) by a *set* of *Q*-functions (*P*-functions). When the set is single-membered, the credal state (confirmational commitment) will be indistinguishable in all relevant respects from a strict bayesian credal state (confirmational commitment).

On this view, if X starts at t with the precise (i.e., single-membered) confirmational commitment C_1, he can then shift to a confirmational commitment that has as members all the *P*-functions in C_1 as well as the *P*-functions in some other confirmational commitment C_2. (As the technical formulation will indicate, other *P*-functions will be members of C_3 as well.)

C_3 will be "weaker" than C_1 or C_2 in that it will allow more *P*-functions to be "permissible" than either of the other two confirmational commitments alone does. It will allow as permissible all *P*-functions recognized as such according to C_1 and according to C_2. In this sense, the shift to C_3 will beg no questions as to the permissibility of the *P*-functions in the other two confirmational commitments.

Of course, the notion of a permissible *P*-function (and the correlative notion of a permissible *Q*-function according to a credal state) require elucidation. I shall offer only an indirect clarification. The account of rational credence (and confirmational commitment) based on the new proposal will be supplemented by criteria for rational choice which indicate how permissibility determines the admissibility of options. By indicating the connections between permissibility and rational choice, permissibility will have been characterized indirectly.

5. To simplify the technical details, I shall restrict the discussion to characterizing credal states and confirmational commitments for sentences in a given language L which belong to a set M generated as follows: Let $h_1, h_2, \ldots, h_n$ be a finite set of sentences in L all consistent with the urcorpus UK for L and such that UK logically implies the truth of at least and at most one h_i. M is the set of sentences in L which are equivalent, given UK, to a disjunction of zero or more distinct h_is. (A disjunction of zero h_is is, as usual, a sentence inconsistent with UK.)

With this understanding, X's credal state at t will be a set $B_{X,t}$ of functions $Q(x;y)$ where the sentences substituted for "x" are in M and the sentences substituted for "y" are in M and are consistent with $K_{X,t}$. When the sentence substituted for "y" is a member of $K_{X,t}$, I shall write $Q(x) = Q(x;y)$.

The set $B_{X,t}$ must satisfy the following three conditions:

1. *Nonemptiness:* $B_{X,t}$ is nonempty.
2. *Convexity:* Let $B_{e,X,t}$ be the set of functions of the form $Q_e(x) = Q(x;e)$ for fixed e in M consistent with $K_{X,t}$ and Q in $B_{X,t}$. $B_{e,X,t}$ is a convex set for every e consistent with $K_{X,t}$ – i.e., every weighted average of finitely many Q_e in B_e in $B_{e,X,t}$ is also in $B_{e,X,t}$.
3. *Coherence:* Every Q-function in $B_{X,t}$ is a finitely additive probability measure on M where $Q(h;e) = 1$ if $K,e \vdash h$ and $Q(h';e') = Q(h;e)$ if $K \vdash e = e'$ and $h = h'$.

Every Q-function in $B_{X,t}$ is "permissible" according to $B_{X,t}$.

As before, X's confirmational commitment $C_{X,t}(K)$ is a function from feasible corpora of knowledge to potential credal states that X at t considers to be the credal states he should adopt were he to adopt K as his corpus of knowledge. The value of the function for given K, therefore, is a nonempty, convex set of Q-functions relative to K. $C_{X,t}(UK) = C_{X,t}$ is, therefore, a nonempty convex set of P-functions. The principle of confirmational conditionalization introduced previously must now be modified to conform to the new characterization of confirmational commitments and credal states:

Confirmational conditionalization: Let K be obtained from UK by adding e (consistent with UK) to UK and forming the deductive closure. If Q is a member of $C_{X,t}(K)$, there is a P in $C_{X,t}(UK)$ such that $Q(h;f) = P(h;f\&e)$ and for every P in $C_{X,t}(UK)$, there is a Q in $C_{X,t}(K)$ such that $Q(h;f) = P(h;f\&e)$.

$B_{X,t}$ can be determined, as before, as follows:

Total knowledge: $B_{X,t} = C_{X,t}(K_{X,t})$.

Thus, X's confirmational commitment is defined by specifying the value of $C_{X,t}(UK)$.

A strict bayesian confirmational commitment, of course, allows a single P-function to be uniquely permissible. However, confirmational commitments are possible which contain more than one P-function. In general, I shall say that one confirmational commitment is stronger than another if the set of its P-functions is a subset of the set of P-functions in the other commitment.

On this view, the weakest confirmational commitment possible is that which contains all the P-functions that meet the requirements of inductive logic. I shall continue to follow Carnap in understanding inductive logic to be a system of principles that impose constraints on probability functions eligible for membership in confirmational commitments.

In contrast the strongest confirmational commitment would be the empty one – which is inconsistent with our first requirement of nonemptiness. A strongest "consistent" confirmational commitment is single-membered.

We can, by the way, extend the notion of a confirmational commitment so as to define it for an inconsistent corpus. We can require that $C_{X,t}(K)$, where K is inconsistent, be empty. This means that our previous requirement that a credal state be nonempty is to be restricted to cases where K is consistent. Thus, X might adopt a consistent confirmational commitment (i.e., one that is nonempty). Yet, if he should, unfortunately, endorse an inconsistent K, his credal state should be empty.

As noted previously, strict bayesians have differed among themselves as to what constitutes a complete system of principles of inductive logic. These differences persist on the view I am now proposing. They may be viewed, however, in a new light. The disagreements over inductive logic turn out to be disagreements over what constitutes the "weakest" possible confirmational commitment – which I shall call "$CIL(UK)$."

"Coherentists" like de Finetti and Savage claim that the principle of coherence constitutes a complete inductive logic. On their view, $CIL(UK)$ is the set of all P-functions obeying the calculus of probabilities defined over M.

Some authors are prepared to add a further principle to the principle of coherence. This principle determines permissible Q-values for hypotheses about the outcome of a specific experiment on a chance device, given suitable knowledge about the experiment to be performed and the chances of possible outcomes of experiments of that type.

There is considerable controversy concerning the formulation of such a principle of "direct inference." In large measure, the controversy reflects disagreements over the interpretation of "chance" or "statistical probability," concerning the so-called problem of the reference class and random sampling. Indeed, the reason coherentists do not endorse a principle linking objective chance with credence is that they either deny the intelligibility of the notion of objective chance or argue in favor of dispensing with that notion.

Setting these controversies to one side, I shall call anyone who holds that a complete inductive logic consists of the coherence principle and an additional principle of direct inference from knowledge of chance to outcomes of random experiments an "objectivist."

There are many authors who are neither coherentists nor objectivists because they wish to supplement the principles of coherence and direct inference with additional principles. Some follow J. M. Keynes, Jeffreys, and Carnap in adding principles of symmetry of various kinds. Others, like I. Hacking (1965, p. 135), introduce principles of irrelevance or other criteria which attempt to utilize knowledge about chances in a manner different from that employed in direct inference. Approaches of this sort stem by and large from the work of R. A. Fisher. I lack a good tag for this somewhat heterogeneous group of viewpoints. They all agree, however, in denying that objectivist inductive logic is a complete inductive logic.

Attempting to classify the views of historically given authors concerning inductive logic is fraught with risk. I shall not undertake a tedious and thankless task of textual analysis in the vain hope of convincing the reader that many eminent authors have been committed to an inductive logic whether they have said so or not. Yet much critical insight into controversies concerning probability, induction, and statistical inference can be obtained by reading the parties to the discussion as if they were committed to some form of inductive logic. If I am right, far from being a dead issue, inductive logic remains very much alive and debated (at least implicitly) not only by bayesians of the Keynes-Jeffreys-Carnap persuasion but by objectivists to whose number I think belong J. Neyman, H. Reichenbach, and authors such as H. Kyburg and I. Hacking (in his first book) who are associated in different ways with the tradition of R. A. Fisher.

Assuming, for the sake of the argument, that the debate concerning what constitutes a complete set of principles of inductive logic is settled (I, for one, would defend and will defend elsewhere adopting a variant of an objectivist inductive logic), there is yet another dimension to debates among students of probability, induction, and statistical inference.

Some authors seem to endorse the view that a rational agent should adopt the weakest confirmational commitment, *CIL,* consonant with inductive logic and hold it fast. They are, in effect, advocating confirmational tenacity. They do so, however, on the grounds that one should not venture to endorse a confirmational commitment stronger than the weakest allowed by inductive logic. (Their view is analogous to one that would require adopting the weakest corpus of knowledge *UK* and holding it fast.) I shall call advocates of such a view "necessitarians."

Again, classifying historically given authors is a risky business. However, Keynes, Jeffreys, and Carnap (in his early work) seem to be clear examples of necessitarians. What is more interesting is the implication that anyone is a necessitarian who insists that the only conditions under which a numerically precise probability can be assigned to a statement (other than a statement that is certainly true or false) are those derivable via direct inference from knowledge of chances. Such authors, on my view, are committed to saying that, when numerical probabilities are not assignable in this way, any numerical value is a permissible assignment provided that it is derived from *Q*-functions allowed by inductive logic.

To illustrate, suppose that *X* knows that a given coin has a .4 or a .6 chance of landing heads on a toss. Let h_1 be the first hypothesis that the chance is .4, and h_2 the second hypothesis. Let g be the hypothesis that the coin will land heads on the next toss. By direct inference, every permissible *Q*-function in *X*'s credal state must be such that $Q(g;h_1) = .4$ and $Q(g;h_2) = .6$. By coherence, every *Q*-function in his credal state must be such that $Q(h_2) = 1 - Q(h_1)$, where $Q(h_1)$ is some real number between 0 and 1 and $Q(g) = Q(g;h_1)Q(h_1) + Q(g;h_2)Q(h_2) = .4Q(h_1) + .6(1 - Q(h_1))$.

According to the authors I have in mind, there is no unique numerical value that a rational *X* should adopt as uniquely permissible for $Q(h_1)$. As I am interpreting such authors as Kyburg, Neyman, Reichenbach, and Salmon, they mean to say that *X*'s credal state should consist of all *Q*-functions meeting the conditions specified. The upshot is that the set of permissible *Q*-values for g should consist of all *Q*-values in the interval from .4 to .6. If I am reading them right, they endorse an objectivist logic and, at the same time, insist that *X* should adopt *CIL* as his confirmational commitment. They are "objectivist necessitarians."

The early Carnap, as noted previously, had hoped to identify an inductive logic that singled out a unique *P*-function as eligible for membership in confirmational commitments. Had his hope been realized, a rational agent would perforce have had to be a necessitarian. The weakest confirmational commitment would have been the strongest consistent one as well. Confirmational tenacity would have been necessitated by the principles of inductive logic.

But if Carnap's program is abandoned, necessitarianism is by no means the only response that one can make. Indeed, it seems to be of doubtful tenability, if for no other reason than that credal states formed on a necessitarian basis seem to be too weak for use in practical decision making or statistical inference. (Many objectivist necessitarians seem to deny this; but the matter is much too complicated to discuss here.)

Personalists, such as de Finetti and Savage, abandon necessitarianism but continue to endorse confirmational tenacity - at least during normal periods free from revolutionary stress. It is this position that I contended earlier leads to dogmatism or capriciousness with respect to confirmational commitment.

The view I favor is *revisionism.* This view agrees with the personalist position in allowing rational men to adopt confirmational commitments stronger than *CIL.* It insists, however, that such commitments are open to revision. It sees as a fundamental epistemological problem the task of providing an account of the conditions under which such revision is appropriate and criteria for evaluating proposed changes in confirmational commitment on those occasions when such shifts are needed.

I shall not offer an account of the revision of confirmational commitments. The point I wish to emphasize here is that, once one abandons the strict bayesian approach to credal rationality and allows credal states to contain more than one permissible *Q*-function in the manner I am suggesting, the revisionist position can be seriously entertained. The strict bayesian view precludes it and leaves us with the dubious alternatives of necessitarianism and personalism. By relaxing the strict bayesian requirements on credal rationality, we can at least ask a question about revision which could not be asked before.

6. According to the approach I am proposing, *X*'s credal state at *t* is characterized by a set of *Q*-functions defined over sentences in a set *M.* Such a representation describes *X*'s credal state globally. Nothing has been said thus far as to how individual sentences in *M* are to be assigned grades of credence or how the degrees of credence assigned to two or more sentences are to be compared with one another. The following definitions seem to qualify for this purpose:

Definition 1: $Cr_{X,t}(h;e)$ is the set of real numbers r such that there is a Q-function in $B_{X,t}$ according to which $Q(h;e) = r$.

Definition 2: $c_{X,t}(h;e)$ is the set of real numbers r such that there is a P-function in $C_{X,t}$ according to which $P(h;e) = r$.

In virtue of the convexity requirement, both the credence function $Cr_{X,t}(h;e)$ and the confirmation function $c_{X,t}(h;e)$ will take sets of values that are subintervals of the unit line - i.e., the interval from 0 to 1. The lower and upper bounds of such intervals have properties which have been investigated by I. J. Good (1962), C. A. B. Smith (1961), and A. P. Dempster (1967).

A partial ordering with respect to comparative credence or with respect to comparative confirmation can be defined as follows:

Definition 3: $(h;e) \overset{Cr_{X,t}}{\leq} (h';f')$ if and only if, for every Q-function in $B_{X,t}$, $Q(h;e) \leq Q(h';e')$.

Definition 4: $(h;e) \overset{c_{X,t}}{\leq} (h';e')$ if and only if, for every P-function in $C_{X,t}$, $P(h;e) \leq P(h';e')$.

The partial orderings induced by credal states and confirmational commitments conform to the requirements of B. O. Koopman's (1940) axioms for comparative probability. Koopman pioneered in efforts to relax the stringent requirements imposed by bayesians on rational credence. Within the framework of his system, he was able not only to specify conditions of rational comparative probability judgment but to identify ways of generating interval-valued credence functions.

According to Koopman's approach, however, any two credal states (confirmational commitments) represented by the same partial ordering of the elements of M are indistinguishable. My proposal allows for important differences. Several distinct convex sets of probability distributions over the elements of M can induce the same partial ordering on the elements of M according to definitions 3 and 4.

Dempster (1967), Good (1962), Kyburg (1961), Smith (1961), and F. Schick (1958) have all proposed modifying bayesian doctrine by allowing credal states and confirmational commitments to be represented by interval-valued probability functions. Good, Smith, and Dempster have also explored the representation of credal states defined by interval-valued credence functions by means of sets of probability measures. Smith and Dempster explicitly consider convex sets of measures. Nonetheless, all these authors, including Dempster and Smith, seem to regard credal states (and confirmational commitments) represented by the same interval-valued function as indistinguishable. In contrast, my proposal recognizes credal states as different even though they generate the identical interval-valued function - provided they are different convex sets of Q-functions.[2]

Thus, the chief difference between my proposal and other efforts to come to grips with "indeterminate" probability judgments is that my proposal recognizes significant differences between credal states (confirmational commitments) where other proposals recognize none. Is this a virtue, or are the fine distinctions allowed by my proposal so much excess conceptual baggage?

I think that the distinctions between credal states recognized by the proposals introduced here are significant. Agents X and Y, who confront the same set of feasible options and evaluate the possible consequences in the same way, may, nonetheless, be obliged as rational agents to choose different options if their credal states are different, even though their credal states define the same interval-valued credence function. That is to say, according to the decision theory that supplements the account of rational credence just introduced, differences in credal states recognized by my theory, but not by Dempster's or Smith's, do warrant different choices in otherwise similar contexts of choice.

To explain this claim, we must turn to a consideration of rational choice. We would have to do so anyhow. One of the demands that can fairly be made of those who propose theories that rival bayesianism is that they furnish answers not only to the problems of rational credence and revision but to the questions about rational choice. Furthermore, the motivation for requiring credal states to be nonempty, convex sets of probability measures and the explanation of the notion of a permissible Q-function are best understood within the context of an account of rational choice. For all these reasons, therefore, it is time to discuss rational choice.

7. Consider, once more, a situation where X faces a decision problem of the type described in section 1. No longer, however, will it be supposed that X's credal state for the "states of nature" $h_1, h_2, \ldots, h_n$ and for the possible consequences $o_{i1}, o_{i2}, \ldots, o_{im}$ conditional on X choosing A_i

2 The difference between my approach and Smith's was drawn to my attention by Howard Stein. To all intents and purposes, both Dempster and Smith represent credal states by the largest convex sets that generate the interval-valued functions characterizing those credal states. Dempster (1967, 332/3) is actually more restrictive than Smith. Dempster, by the way, wrongly attributes to Smith the position I adopt. To my knowledge, Dempster is the first to consider this position in print - even if only to misattribute it to Smith.

are representable by a single Q-function. Instead, the credal state will be required only to be a nonempty convex set of Q-functions.[3]

Although I have not focused attention here on the dubiety of requiring X's evaluations of the o_{ij}'s to be representable by a utility function unique up to a positive affine transformation, I do believe that rational men can have indeterminate preferences and will, for the sake of generality, relax the bayesian requirement as follows: X's system of evaluations of the possible consequences of the feasible options is to be represented by a set G of "permissible" u-functions defined over the o_{ij}'s which is (a) nonempty, (b) convex, and such that all positive affine transformations of u-functions in G are also in G. A strict bayesian G is, in effect, such that all u-functions in it are positive affine transformations of one another. It is this latter requirement that I am abandoning.

In those situations where X satisfies strict bayesian conditions so that his credal state contains only a single Q-function and G contains all and only those u-functions which are positive affine transformations of some specific u-function u_1, an admissible option A_1, is, according to the principle of maximizing expected utility, an option that bears maximum expected utility

$$E(A_i) = \sum_{i=1}^{m} Q(h_i)u_1(o_{ij})$$

Notice that, if any positive affine transformation of u_1 is substituted for u_1 in the computation of expected utility, the ranking of options with respect to expected utility remains unaltered. Hence we can say that, according to strict bayesians, an option is admissible if it bears maximum expected utility relative to the uniquely permissible Q-function and to any of the permissible Q-functions and to any of the permissible u-functions in G (all of which are positive affine transformations of u_1).

There is an obvious generalization of this idea applicable to situations

3 I am assuming here that the option chosen is confirmationally irrelevant to the states of nature. Let B_{A_i} be the credal state B ($= B_{X,t}$), restricted to the permissible functions of the form $Q(x;A_i$ is chosen) and where the values of x are boolean combinations of hypotheses specifying the states of nature of the form h_j and let B_t be the credal state B restricted to the permissible functions of the form $Q(x;t) = Q(x)$ where t is entailed by $K_{X,t}$. The choosing of A_i is confirmationally irrelevant in the weak sense to the h_j's if and only if the set $B_{A_i} = B_t$. This notion of irrelevance does not imply that $Q(hj;A_i) = Q(hj)$ for every permissible Q-function in B. I call this latter condition "strong confirmational irrelevance" whereas the former is "weak confirmational irrelevance." When strong confirmational irrelevance is satisfied, the E-admissible options can be determined using the unconditional Q-functions in B_t to compute expected utilities. For further discussion of confirmational irrelevance, see I. Levi (1980, pp. 225–233, and 1978, pp. 263–273). This note represents a revision of footnote 12 in the original version of this paper.

where $B_{X,t}$ contains more than one permissible Q-function and G contains u-functions that are not positive affine transformations of one another. I shall say that A_i is *E-admissible* if and only if there is at least one Q-function in $B_{X,t}$ and one u-function in G such that $E(A_i)$ defined relative to that Q-function and u-function is a maximum among all the feasible options. The generalization I propose is the following:

E-admissibility: All admissible options are *E-admissible.*

The principle of *E*-admissibility is by no means novel. I. J. Good (1952, p. 114), for example, endorsed it at one time. Indeed, Good went further than this. He endorsed the converse principle that all *E*-admissible options are admissible as well.

I disagree with Good's view on this. When *X*'s credal state and goals select more than one option as *E*-admissible, there may be and sometimes are considerations other than *E*-admissibility which *X*, as a rational agent, should employ in choosing between them.

There are occasions where *X* identifies two or more options as *E*-admissible and where, in addition, he has the opportunity to defer decision between them. If that opportunity is itself *E*-admissible, he should as a rational agent "keep his options open." Notice that in making this claim I am not saying that the option of deferring choice between the other *E*-admissible options is "better" than the other *E*-admissible options relative to *X*'s credence and values and the assessments of expected utility based thereon. In general, *E*-admissible options will not be comparable with respect to expected utility (although sometimes they will be). The injunction to keep one's options open is a criterion of choice that is based not on appraisals of expected utility but on the "option-preserving" features of options. Deferring choice is better than the other *E*-admissible options in this respect, but not with respect to expected utility.

Thus, a *P*-admissible option is an option that is (a) *E*-admissible and (b) "best" with respect to *E*-admissible option preservation among all *E*-admissible options. I shall not attempt to provide an adequate explication of clause (b) here. In the subsequent discussion, I shall consider situations where there are no opportunities to defer choice. Nonetheless, it is important to notice that, given a suitably formulated surrogate for (b), the following principle holds:

P-admissibility: All admissible options are *P*-admissible.

My disagreement with Good goes still further than this; for I reject not only the converse of *E*-admissibility but that of *P*-admissibility as well.

To illustrate, consider a situation that satisfies strict bayesian requirements. *X* knows that a coin with a .5 chance of landing heads is to be tossed once. *g* is the hypothesis that the coin will land heads. Under the circumstances, we might say that *X*'s credal state is such that all permissible *Q*-functions assign *g* the value $Q(g) = .5$. Suppose that *X* is offered a gamble on *g* where *X* gains a dollar if *g* is true and loses one if *g* is false. (I shall assume that *X* has neither a taste for nor an aversion to gambling and that, for such small sums, money is linear with utility.) He has two options: to accept the gamble and to reject it. If he rejects it, he neither gains nor loses.

Under the circumstances described, the principle of maximizing expected utility may be invoked. It indicates that both options are optimal and, hence, in my terms *E*-admissible. Since there are no opportunities for delaying choice, both options (on a suitably formulated version of *P*-admissibility) become *P*-admissible.

Bayesians - and Good would agree with this - tend to hold that rational *X* is free to choose either way. Not only are both options *E*-admissible. They are both admissible. Yet, in my opinion, rational *X* should refuse the gamble. The reason is not that refusal is better in the sense that it has higher expected utility than accepting the gamble. The options come out equal on this kind of appraisal. Refusing the gamble is "better," however, with respect to the security against loss it furnishes *X*. If *X* refuses the gamble, he loses nothing. If he accepts the gamble, he might lose something. This appeal to security does not carry weight, in my opinion, when accepting the gamble bears higher expected utility than refusing it. However, in that absurdly hypothetical situation where they bear precisely the same expected utility, the question of security does become critical.

These considerations can be brought to bear on the more general situation where two or more options are *E*-admissible (even though they are not equal with respect to expected utility) and where the principle of *P*-admissibility does not weed out any options.

An *S*-admissible option (i.e., option admissible with respect to security) is an option that is *P*-admissible and such that there is a permissible *u*-function in *G* relative to which the minimum *u*-value assigned a possible consequence o_{ij} of option A_i is a maximum among all *P*-admissible options.[4]

4 Maximin principles in particular and criteria of *S*-admissibility in general suffer from a serious ambiguity. For example, the set of possible consequences of a "mixed act" constructed by choosing between "pure options" A_i and A_j with the aid of a chance device with known chance probability of selecting one or the other option may be

S-admissibility: All admissible options are *S*-admissible.

I cannot think of additional criteria for admissibility which seem adequate. (But then I have no precise conditions of adequacy.) I think, perhaps, we should keep an open mind on this matter. Nonetheless, for the present, I shall tentatively assume that the converse of *S*-admissibility holds. This assumption will not alter the main course of the subsequent argument.

Even without detailed exploration of the ramifications of this decision theory, some of its main features are immediately apparent. It conforms to the strict bayesian injunction to maximize expected utility in those situations where X has a precise credal state and G contains u-functions that are all linear transformations of one another. In this sense, bayesian decision theory is a special case of mine.

Similarly, the proposed decision theory identifies situations where the well-known maximin criterion is applied legitimately. Customarily maximin is used to select that option from among all the *feasible* options which maximizes the minimum gain. This recommendation is legitimate, according to my theory, provided (1) G contains all and only u-functions that are positive affine transformations of one another, and (2) all feasible options are *P*-admissible. But even if condition (1) is satisfied, it can be the case that the maximin solution from among all the feasible options is not itself *E*-admissible and so cannot be considered to be *S*-admissible.

Finally, my proposal is able to discriminate between and cover a wider variety of situations where neither maximizing expected utility nor max-

construed as the set of possible consequences of either A_i or A_j. According to this conception of possible consequences of the mixed act, the security level for a given u-function is the lowest of the security levels belonging to A_i and A_j. Hence, one cannot raise the security level by taking a mixture of two options. Wald, von Neumann, and Morgenstern proceed differently. They take a possible consequence of the mixed option to be uniquely determined by the "state of nature." Given the state of nature, the mixed option is equivalent to a lottery whose value is equal to its expected utility given the state of nature. Such a lottery is then the "consequence" of the mixed option given the state of nature. The set of consequences so determined by the possible states of nature are the possible consequences of the mixed option for the purposes of determining security. Security levels for mixed options may then be determined differently and can, indeed, be higher than the security levels of the pure options involved in the mixture. In an earlier version of this footnote, I favored construing *S*-admissibility using the first conception of a security level. In later publications, I abandoned this view. I now think that the way in which a decision maker individuates possible consequences of options for the purpose of identifying security levels is up to the agent and reflects a feature of the agent's value commitments which ought not to be dictated to the agent by principles of "thin" rationality. For further discussion of this point, see I. Levi (1980, pp. 156–163). In any case, as was pointed out in the original version of this note, mixtures of *E*-admissible options are not always *E*-admissible. In this paper, I leave mixed options out of account.

imining can be invoked with much plausibility. Moreover, it does so with the aid of a unified system of criteria of rational credence and rational choice. Thus, it does offer answers to just those questions which Bayesian theory purports to solve. Moreover, it escapes the bayesian commitment to the dubious doctrines of necessitarianism or personalism.

8. Some elementary properties of credal states as nonempty convex sets will be illustrated and explained by applying the decision theory just outlined to simple gambling situations. Suppose X knows that a coin is to be tossed and has either a .4 or .6 chance of landing heads. g is the hypothesis that the coin will land heads. I shall suppose that X has neither a taste for nor an aversion to gambling and that X's values are such that G is a set of u-functions that are linear transformations of the monetary payoffs of the gambles to be considered.

Case 1: X is offered a gamble on a take-it-or-leave-it basis where he receives $S-P$ dollars if g is true and loses P dollars if g is false. (Both S and P are positive.)

Case 2: X is offered a gamble on a take-it-or-leave-it basis where he loses P dollars if g is true and receives $S-P$ dollars if g is false. (S and P have the same values as in case 1.)

h_1 is the hypothesis that the chance of heads is .4, and h_2 is the hypothesis that the chance of heads is .6. By the reasoning of page 130, every permissible Q-function in X's credal state should be such that $Q(g) = .4Q(h_1) + .6[1 - Q(h_1)]$.

According to strict bayesians, X should, therefore, adopt a credal state that selects a single such Q-function as permissible. This can be done by selecting a single value for $Q(h_1)$. If that value is r; $Q(g) = .4r + .6(1-r) = .6 - .2r$.

Hence, the bayesian will find that accepting the case 1 gamble is uniquely admissible if and only if $Q(g) > P/S$, and will find accepting the case 2 gamble uniquely admissible if and only if $Q(\sim g) > P/S$. (Otherwise rejecting the gamble for the appropriate case is uniquely admissible, assuming that ties in expected utility are settled in favor of rejection.) Hence, if P/S is less than .5, a bayesian must preclude the possibility of accepting the gamble being inadmissible both in case 1 and in case 2.

Suppose, however, that $Cr_{X,t}(h_1)$ takes a nondegenerate interval as a value. For simplicity, let that interval be [0,1]. The set of permissible Q-values for g must be all values of $.6 - .2r$ where r takes any value from 0 to 1. Hence, $Cr_{X,t}(g) = [.4, .6]$.

Under these conditions, my proposal holds that, when P/S falls in the interval from .4 to .6, both options are E-admissible (and P-admissible) in case 1. The same is true in case 2. But in both case 1 and case 2, rejecting the gamble is uniquely S-admissible. Hence, in both cases, X should reject the gamble. *This is true even when P/S is less than .5.* In this case, my proposal allows a rational agent a system of choices that a strict bayesian would forbid. In adopting this position, I am following the analysis advocated by C. A. B. Smith for handling pairwise choices between accepting and rejecting gambles. Smith's procedure, in brief, is to characterize X's degree of credence for g by a pair of numbers (the "lower pignic probability" and the "upper pignic probability" for g) as follows: The lower pignic probability $\underline{s}$ represents the least upper bound of betting quotients P/S for which X is prepared to accept gambles on g for positive S. The upper pignic probability $\bar{s}$ for g is $1-t$, where t is the least upper bound of betting quotients P/S for which X is prepared to accept gambles on $\sim g$ for positive S. Smith requires that $\underline{s} \leq \bar{s}$, but does not insist on equality as bayesians do. Given Smith's definitions of upper and lower pignic probabilities, it should be fairly clear that, in case 1 and case 2 where $Cr_{X,t}(g) = [.4, .6]$, Smith's analysis and mine coincide.[5]

Before leaving cases 1 and 2, it should be noted that, if X's credal state were empty, no option in case 1 would be admissible and no option in case 2 would be admissible either. If X is confronted with a case 1 predicament and an empty credal state, he would be constrained to act and yet as a rational agent enjoined not to act. The untenability of this result is to be blamed on adopting an empty credal state. Only when X's corpus is inconsistent, should a rational agent have an empty credal state. But, of course, if X finds his corpus inconsistent, he should contract to a consistent one.

Case 3: A_1 is accepting both the case 1 and the case 2 gamble jointly with a net payoff if g is true or false of $S-2P$.

5 Smith (1961, pp. 3–5, 6–7). The agreement applies only to pairwise choices where one option is a gamble in which there are two possible payoffs and the other is refusing to gamble with 0 gain and 0 loss. In this kind of situation, it is clear that Smith endorses the principle of E-admissibility, but not its converse. However, in the later sections of his paper where Smith considers decision problems with three or more options or where the possible consequences of an option to be considered are greater than 2, Smith seems (but I am not clear about this) to endorse the converse of the principle of E-admissibility - counter to the analysis on the basis of which he defines lower and upper pignic probabilities. Thus, it seems to me that either Smith has contradicted himself or (as is more likely) he simply does not have a general theory of rational choice. The latter sections of the paper may then be read as interesting explorations of technical matters pertaining to the construction of such a theory, but not as actually advocating the converse of E-admissibility. At any rate, since it is the theory Smith propounds in the first part of his seminal essay which interests me, I shall interpret him in the subsequent discussion as having no general theory of rational choice beyond that governing the simple gambling situations just described.

This is an example of decision making under certainty. Everyone agrees that if P is greater than $2S$ the gamble should be rejected; for it leads to certain loss. If P is less than $2S$ X should accept the gamble; for it leads to a certain gain. These results, by the way, are implied by the criteria proposed here as well as by the strict bayesian view.

Strict bayesians often defend requiring that Q-functions conform to the requirements of the calculus of probabilities by an appeal to the fact that, when credal states contain but a single Q-function, a necessary and sufficient condition for having credal states that do not license sure losses (Dutch books) is having a Q-function obeying the calculus of probabilities. The arguments also support the conclusion that, even when more than one Q-function is permissible according to a credal state, if all permissible Q-functions obey the coherence principle, no Dutch book can become E-admissible and, hence, admissible.

Case 4: B_1 is accepting the case 1 gamble, B_2 is accepting the case 2 gamble, and B_3 is rejecting both gambles.

Let the credal state be such that all values between 0 and 1 are permissible Q-values for h_i and, hence, all values between .4 and .6 are permissible for g.

If P/S is greater than .6, B_3 is uniquely E-admissible and, hence, admissible. If P/S is less than .4, B_3 is E-inadmissible. The other two options are E-admissible and admissible.

If P/S is greater than or equal to .4 and less than .5, B_3 remains inadmissible and the other two admissible.

If P/S is greater than or equal to .5 and less than .6, all three options are E-admissible; but B_3 is uniquely S-admissible. Hence, B_3 should be chosen when P/S is greater than or equal to .5.

Three comments are worth making about these results.

1. I am not sure what analysis Smith would propose of situations like case 4. At any rate, his theory does not seem to cover it (but see footnote 5).
2. When P/S is between .4 and .5, my theory recommends rejecting the gamble in case 1, rejecting the gamble in case 2, and yet recommends accepting one or the other of these gambles in case 4. This violates the so-called principle of independence of irrelevant alternatives.[6]

6 See Luce and Raiffa (1958, pp. 288–289). Because the analysis offered by Smith and me for cases 1 and 2 seems perfectly appropriate and the analysis for case 4 also appears impeccable, I conclude that there is something wrong with the principle of independence of irrelevant alternatives.

3. If the convexity requirement for credal states were violated by removing as permissible values for g all values from $(S-P)/S$ to P/S, where P/S is greater than .5 and less than .6, but leaving all other values from .4 to .6, then - counter to the analysis given previously, B_3 would not be E-admissible in case 4. The peculiarity of that result is that B_1 is E-admissible because, for permissible Q-values from .6 down to P/S, it bears maximum expected utility, with B_3 a close second. B_2 is E-admissible because, for Q-values from .4 to $(S-P)/S$, B_2 is optimal, with B_3 again a close second. If the values between $(S-P)/S$ and P/S are also permissible, B_3 is E-admissible because it is optimal for those values. To eliminate such intermediate values and allow the surrounding values to retain their permissibility seems objectionable. Convexity guarantees against this.

Case 5: X is offered a gamble on a take-it-or-leave-it basis in which he wins 15 cents if f_1 is true, loses 30 cents if f_2 is true, and wins 40 cents if f_3 is true.

Suppose X's corpus of knowledge contains the following information:

Situation a: X knows that the ratios of red, white, and blue balls in the urn are either (i) 1/8, 3/8, 4/8 respectively; (ii) 1/8, 4/8, 3/8; (iii) 2/8, 4/8, 2/8; or (iv) 4/8, 3/8, 1/8.

X's credal state for the f_i's is determined by his credal state for the four hypotheses about the contents of the urn according to a more complex variant of the arguments used to obtain credence values for g in the first four cases. If we allow all Q-functions compatible with inductive logic or an objectivist kind to be permissible, X's credal state for the f_i's is the convex set of all weighted averages of the four triples of ratios. $Cr_{X,t}(f_1) = (1/8, 4/8)$, $Cr_{X,t}(f_2) = (3/8, 4/8)$, and $Cr_{X,t}(f_3) = (1/8, 4/8)$. Both accepting and rejecting the gamble are E-admissible. Rejecting the gamble, however, is uniquely S-admissible. X should reject the gamble.

Situation b: X knows that the ratios of red, white, and blue balls is correctly described by (i), (ii), or (iv), but not by (iii). Calculation reveals that the interval-valued credence function is the same as in situation *a*. Yet it can be shown that

A hint as to the source of the trouble can be obtained by noting that if "E-admissible" is substituted for "optimal" in the various formulations of the principle cited by Luce and Raiffa, p. 289, the principle of independence of irrelevant alternatives stands. The principle fails because S-admissibility is used to supplement E-admissibility in weeding out options from the admissible set.

Mention should be made in passing that even when "E-admissible" is substituted for "optimal" in Axiom 9 of Luce and Raiffa, p. 292, the axiom is falsified. Thus, when $.5 \leq P/S \leq .6$ in case 4, all three options are E-admissible, yet some mixtures of B_1 and B_2 will not be.

accepting the gamble is uniquely E-admissible and, hence, admissible. X should accept the gamble.

Now we can imagine situations that are related as *a* and *b* are to one another except that the credal states do not reflect differences in statistical knowledge. Then, from the point of view of Dempster and Smith, the credal states would be indistinguishable. Because the set of permissible Q-distributions over the f_i's would remain different for situations *a* and *b*, my view would recognize differences and recommend different choices. If the answer to the problem of rational choice proposed here is acceptable, the capacity of the account of credal rationality to make fine distinctions is a virtue rather than a gratuitous piece of pedantry.

The point has its ramifications for an account of the improvement of confirmational commitments; the variety of discriminations that can be made between confirmational commitments generates a variety of potential shifts in confirmational commitments subject to critical review. For intervalists, a shift from situation *a* to *b* is no shift at all. On the view proposed here, it is significant.

The examples used in this section may be used to illustrate one final point. The objective or statistical or chance probability distributions figuring in chance statements can be viewed as assumptions or hypotheses. Probabilities in this sense can be unknown. We can talk of a set of simple or precise chance distributions among which X suspends judgment. Such *possible* probability distributions represent hypotheses which are possibly true and which are themselves objects of appraisal with respect to credal probability. *Permissible* probability distributions which, in our examples, are defined over such *possible* probability distributions (like the hypotheses h_1 and h_2 of cases 1, 2, 3, and 4) are not themselves possibly true hypotheses. No probability distributions of a still higher *type* can be defined over them.[7]

7 I mention this because I. J. Good, whose seminal ideas have been an important influence on the proposals made in this essay, confuses permissible with possible probabilities. As a consequence, he introduces a hierarchy of types of probability (Good, 1962, p. 327). For criticism of such views, see Savage (1954, p. 58). In fairness to Good, it should be mentioned that his possible credal probabilities are interpreted not as possibly true statistical hypotheses but as hypotheses entertained by X about his own unknown strictly bayesian credal state. Good is concerned with situations where strict bayesian agents having precise probability judgments cannot identify their credal states before decision and must make choices on the basis of partial information about themselves. [P. C. Fishburn (1964) devotes himself to the same question.] My proposals do not deal with this problem. I reject Good's and Fishburn's view that every rational agent is at bottom a strict bayesian limited only by his lack of self-knowledge, computa-

I have scratched the surface of some of the questions raised by the proposals made in this essay. Much more needs to be done. I do believe, however, that these proposals offer fertile soil for cultivation not only by statisticians and decision theorists but by philosophers interested in what, in my opinion, ought to be the main problem for epistemology – to wit, the improvement (and, hence, revision) of human knowledge and belief.

REFERENCES

Carnap, R. (1962), *Logical Foundations of Probability,* University of Chicago Press, 1950; 2nd ed., 1962.

(1971a), "Inductive logic and rational decisions," in *Studies in Inductive Logic and Probability,* ed. by R. Carnap and R. C. Jeffrey, UCLA Press.

(1971b), "A basic system of inductive logic," in *Studies in Inductive Logic and Probability,* ed. by R. Carnap and R. C. Jeffrey, UCLA Press.

Dempster, A. P. (1967), "Upper and lower probabilities induced by a multivalued mapping," *Annals of Mathematical Statistics* 38, 325-39.

Fishburn, P. C. (1964), *Decision and Value,* Wiley.

Good, I. J. (1952), "Rational decisions," *Journal of the Royal Statistical Society, Ser. B* 14, 107-14.

(1962), "Subjective probability as a measure of a nonmeasurable set," in *Logic, Methodology and Philosophy of Science, Proceedings of the 1960 International Congress,* ed. by E. Nagel, P. Suppes and A. Tarski, Stanford University Press, 319-29.

Hacking, I. (1965), *Logic of Statistical Inference,* Cambridge University Press.

Koopman, B. O. (1940), "The bases of probability," *Bulletin of the American Mathematical Society* 46, 763-74.

Kyburg, H. E. (1961), *Probability and the Logic of Rational Belief,* Wesleyan University Press, 1961.

Levi, I. (1978), "Irrelevance," in *Foundations and Applications of Decision Theory,* vol. 1, ed. by C. A. Hooker, J. J. Leach and E. F. McClennen, Reidel, 263-73.

(1980), *The Enterprise of Knowledge,* MIT Press.

Luce, R. D. and Raiffa, H. (1958), *Games and Decisions,* Wiley.

Savage, L. J. (1954), *The Foundations of Statistics,* Wiley, 2nd rev. edition, Dover, 1972.

tional facility, and memory. To the contrary, I claim that, even without such limitations, rational agents should not have precise bayesian credal states. The difference in problem under consideration and presuppositions about rational agents has substantial technical ramifications which cannot be developed here.

Schick, F. (1958), *Explication and Inductive Logic,* Ph.D. dissertation, Columbia University.

Smith, C. A. B. (1961), "Consistency in statistical inference" (with discussion), *Journal of the Royal Statistical Society, Ser. B* 23, 1-25.

Spielman, S. (1975), "Levi on personalism and revisionism," *Journal of Philosophy* 62, 785-93.

7

Consensus as shared agreement and outcome of inquiry

When two or more agents disagree concerning what is true, concerning how likely it is that a hypothesis is true or concerning how desirable it is that a hypothesis be true, they may, of course, be prepared to rest content with the disagreement and treat each other's views with contemptuous toleration. The mere fact that someone disagrees with one's judgments is insufficient grounds for opening up one's mind. Most epistemologists forget that it is just as urgent a question to determine when we are justified in opening up our minds as it is to determine when we are justified in closing them.

Nonetheless, the context of disagreement sometimes offers good reasons for the participants initiating some sort of investigation to settle their dispute. When this is so, an early step in such a joint effort is to identify those shared agreements which might serve as the noncontroversial basis of subsequent inquiry. Once this is done, investigation may proceed relative to that background of shared agreements according to those methods acknowledged to be appropriate to the problem under consideration.

On this view of disagreement and its resolution through inquiry, the notion of consensus may be used in two quite distinct ways: (a) One can speak of that consensus of the participants at the beginning of inquiry which constitutes the background of shared agreements on which the investigation is initially grounded. (b) Sometimes inquiry of this kind may terminate with a satisfactory conclusion. We may then say that a consensus has been reached as to the outcome of inquiry.

The contrast between consensus as shared agreement and consensus as outcome of inquiry will seem unimportant to someone who denies the feasibility of conducting inquiry into a certain category of problems without begging the issues under dispute. As I understand him, Kuhn thinks that controversies over rival theories in a revolutionary context are of this kind. To shift to a system of shared assumptions and investigate

From *Synthese* 62 (1985), pp. 3–11.

the merits of rival hypotheses is not feasible. The system of shared assumptions, if intelligible at all, is far too weak to form a basis for subsequent inquiry aimed at deciding between rival claims. From such a point of view, there is only one kind of consensus that is of interest: the consensus reached when the revolution is over and one view has emerged the victor to guide inquiry in the subsequent period of normal science. This is consensus neither as shared agreement nor as outcome of inquiry. It is a surrogate for consensus as outcome of inquiry in that it is a settling of the dispute. But in this case, the dispute is settled by revolution, conversion, gestalt switch or, perhaps, some other psychological, social or political process. The outcome is not the product of a genuine inquiry in which pros and cons are weighed from a point of view which begs no questions under dispute.

I do not think that Kuhn or anyone else has made a case supporting this view. The appeal to history so fashionable in this connection often strikes me more as a ritualistic incantation than reference to a substantial argument. I shall take it for granted, without further discussion, that it is desirable wherever feasible to resolve disputes by engaging in inquiry based on shared agreements which beg no controversial issues. When this approach is taken, the difference between consensus as shared agreement and as outcome of inquiry is relevant, important and clear.

I shall suppose that this attitude ought not to be restricted to cases where investigators differ concerning what is to be accepted as true elements of the evolving doctrine but should be applied to disagreements over the assessments of probabilities as well.

Thus, physicians engaged in diagnosing some rather difficult case may differ with one another not in virtue of the fact that one declares it to be true that the patient is suffering from *A* whereas the other physician denies it. Both physicians may be uncertain as to whether the patient suffers from *A* but differ concerning the degree of probability they assign the hypothesis.

In some cases, this disagreement may be reduced to a disagreement concerning the truth of some statistical hypothesis specifying the objective or statistical probability of some indicator of the disease *A* being misleading on the application of the test. But on some occasions, it may turn out that two or more agents cannot identify any truth of the matter over which they disagree. They differ in their credal probability judgments nonetheless.

I suggest that even in such cases, the agents should first identify their shared agreements and modify their credal states so as to restrict themselves to these shared agreements. On this basis, they may then proceed to engage in whatever deliberations may be appropriate to settle their differences. A problem immediately arises if this approach is to be realized. A subjective or credal probability judgment (or a system of such judgments represented by a credal probability distribution over an algebra of propositions) is neither true nor false. (Of course, the claim that some agent *X* at time *t* makes such a judgment has a truth value. The judgment he makes at that time, however, does not.) Hence, the agent cannot suspend judgment as to its truth. He cannot assign the hypothesis that it is true a certain credal probability (Levi, 1977; 1978; 1980, ch. 9.3; 1982).

If that is the case, in what sense can an agent at a time suspend judgment between rival credal probability distributions? The proposal made in Levi (1974) and subsequently elaborated in Levi (1980) was based on the point that credal probability distributions are used in deliberation to calculate expectations and, in particular, the expected values of feasible options in decision problems. To be in suspense between rival distributions is to be in suspense between rival ways of evaluating feasible options with respect to expected value. To be in suspense about the latter is to regard two or more ways of evaluating feasible options as being permissible to use. That is to say, options which come out optimal in a feasible set relative to a permissible evaluation are admissible as far as calculation of expected value is concerned. They are *E*-admissible in my jargon. To suspend judgment between rival probability distributions, therefore, is to regard them all as permissible to use in calculating expectations for the purpose of assessing *E*-admissibility.

As I pointed out in my 1974 paper, this way of thinking of the matter conflicts with strict Bayesian dogma which requires that an ideally rational agent endorse a "credal state" according to which exactly one credal probability distribution is permissible. From this point of view, the agent can never be in suspense between rival distributions. Either he remains loyal to his current probabilistic judgment except for modifications due to updating via Bayes's theorem and conditionalization on the data or he is involved in a probabilistic revolution whereby he becomes converted from one confirmation, credibility function or, in my terminology, confirmational commitment to another.

According to strict Bayesian dogma, therefore, there can be no analogue in contexts of probability judgment of the two senses of consensus I identify. If two or more agents differ in probability judgment, they can all switch either to the distribution adopted by one of them or to some other distribution which is, in a sense, a potential resolution of the conflict between their differing distributions. There is only one kind of consensus to be recognized – namely the resolution of conflict reached through revolution, conversion, voting, bargaining or some other psychological or social process.

In my 1974 paper and subsequent work, I adopted the view that a potential resolution of the conflict between rival credal probability distributions is to be represented by a credal probability distribution which is the weighted average of the distributions in conflict. Hence, the set of all potential resolutions of such a conflict is to be represented by the convex hull (the set of all weighted averages) of the credal distributions initially in contention. My assumption was that this convex set of probability distributions represented the first kind of consensus I regard as important – consensus as shared agreements regarding probability judgment. I took a parallel view concerning disputes over utility judgment as well.

I claim no novelty for introducing the idea of regarding weighted averages of probability distributions or utility distributions as potential resolutions of conflicts of other distributions. Neither do Lehrer and Wagner (1981). At the outset, however, it is important to emphasize this particular shared agreement with their view. We both take for granted that when two or more individuals differ in their credal probability distributions, a potential resolution of the dispute which is itself representable as another uniquely permissible probability distribution should be a weighted average of the distributions in contention.

Lehrer and Wagner seem to endorse, however, the fundamental dogma of strict Bayesianism according to which a rational agent should be committed at all times to a uniquely permissible credal probability distribution. If two or more agents disagree in the distributions they recognize as uniquely permissible, the problem is to find some sort of rational procedure which allows them to reach a consensus of the sort which resolves the dispute.

My objection is precisely that their view prevents one from engaging in a deliberation beginning from a consensus of shared agreements in order to reach a consensus resolving or partially resolving the initial disagreement as the outcome of inquiry.

The upshot is that the Lehrer-Wagner view is capable of recognizing only those changes of probability judgment which have the traits of revolution and conversion – at least in those cases where changes in probability cannot be viewed as updating on the basis of data via conditionalization and Bayes's theorem. They exemplify the view which I described in Levi (1974) as allowing "shifts in confirmational commitment due to conversion under revolutionary stress."

Lehrer and Wagner acknowledge a link with Kuhn, but they apparently have a distaste for revolution. They think that someone who replaces his initial credal distribution by another which is a weighted average of the credal distributions of the individuals in his cognitive community is pooling the information available to these individuals and making their information his own. They claim that their method of reaching consensus "conforms to a central canon of scientific methodology. The principle is that it is irrational to ignore relevant evidence and rational to do the opposite . . ." (Lehrer and Wagner, 1981, pp. 33-34).

Observance of a total relevant knowledge requirement is undoubtedly an ideal of "scientific methodology". But to sustain the claim that their method of reaching consensus conforms to this central canon depends on some assumption concerning how credal probability judgment is to be modified when evidence thus far unrecognized is taken into consideration. Their method of consensus does not go through Bayes's theorem and conditionalization on the evidence or information. So they cannot invoke the authority of Bayesianism to make good their claim to be adhering to a total evidence requirement.

Perhaps Lehrer and Wagner mean to invoke some non-Bayesian principle of updating on the basis of new evidence. They do not tell us what that principle is except that it shows that the method of reaching consensus is a rational way of conforming to the total relevant knowledge requirement.

In my judgment, this way of proceeding is insufficient to distance Lehrer and Wagner from revolution or some other brand of political epistemology.

Consensus is reached according to the Lehrer-Wagner approach by a method which determines it to be a function of the respect with which each member of the cognitive community regards the views of the other members. Such respect is not indexed by the probability accorded the conjecture that some colleague in the community has made a true probability assessment. Credal probability judgments lack truth values. Hence, an agent cannot assign a credal probability value to the hypothesis that some such judgment is true. The sort of respect Lehrer and Wagner have in mind remains obscure to me but it clearly has political overtones. A Lehrer-Wagner consensus is reached by letting each member of the community weight each credal probability judgment made by a member of the community with the respect accorded to him and then taking the weighted average. The resulting probability distribution is somewhat like a "vote" made by the member who computes a weighted average according to his assignments of grades of respect to himself and his colleagues. Each member of the community votes in the same way. The result is a new set of probability distributions - one for each member. The participants vote again in the same manner using the same assessments of respect they did at the previous step. The process is reiterated. Under the technical conditions specified, the reiterated process converges on a definite credal probability distribution which is a weighted average of the initial set of credal probability judgments. That limit is the consensual probability.[1]

The proposal supplies a method for calculating a set of weights representing an appropriate resolution of the conflict in credal probability judgment which occasioned the concern to obtain a resolution. It determines the weights as a function of the respect accorded by each individual to each other individual and to himself and the distributions initially in contention.

Such a procedure is scarcely the sort which would find favor with investigators engaged in inquiry. To the contrary, it is a proposal to settle disputes by algorithm. It urges each party to the dispute to convert from his initial credal distribution to the consensual one not on the basis of the acquisition of new information relative to a background of shared agreement but as the outcome of a rule depending on the collegiality of the members of the community.

To elaborate further on the difference between inquiry oriented views concerning how consensus should be reached and the algorithmic appeal to collegiality favored by Lehrer and Wagner, I shall comment briefly on some remarks made by Robert Laddaga on Lehrer's views (Lehrer, 1978).

Suppose at the outset of the dispute, agent X_1 has a strictly Bayesian credal state such that $P_1(H/E)=P_1(H)$ and X_2 has another such state such that $P_2(H/E)=P_2(H)$. It is well known that a weighted average of P_1 and P_2 assigning positive weights to both P_1 and P_2 will fail, in general, to preserve independence. Let $P_a=aP_1+(1-a)P_2$. In general, $P_a(H/E)$ will differ from $P_a(H)$.[2]

1 It should be emphasized that the normative model for reaching a consensus proposed by Lehrer and Wagner was proposed with a virtually identical construal by M. De Groot in De Groot (1974). Lehrer and Wagner do acknowledge De Groot's contribution; but philosophers interested in the Lehrer-Wagner theory should be aware of a substantial literature on De Groot's ideas in the statistical literature. Berger (1981) provides some helpful bibliography and review of the issues discussed.

2 This point was driven home to me by R. D. Luce in 1975 in a conference in Western Ontario in 1975 while commenting on a paper I delivered on indeterminate probability and irrelevance. Shortly afterwards in 1975, I revised my original paper and it became Levi (1978), appearing as part of the proceedings of that conference in Hooker, Leach, and McLennen (1978).

Laddaga (1977) found this disturbing. I infer from their public responses that Lehrer and Wagner do not.

In one respect, Lehrer and Wagner are right not to be disturbed. Whether one adopts an inquiry oriented model of the resolution of disagreements as I do or the revolutionary conversion model of such resolution favored by Lehrer and Wagner, the kind of consensus which fully resolves disagreement in credal probability judgment is a weighted average of the credal probability judgments in contention. It is not to be expected that the consensus reached as the outcome of inquiry or via algorithmic conversion is going to preserve all assumptions initially shared in common by the participants in the dispute.

On the other hand, according to the inquiry oriented model I favor, at the beginning of inquiry probability judgments should beg no questions in favor of one view in contention or the other while preserving those assumptions which are not in dispute. If everyone agrees that H and E are probabilistically independent, that feature should be represented in the consensus as shared agreement. Lehrer and Wagner cannot, given their vision of how disagreements are to be adjudicated, countenance consensus as shared agreement. In my view, that is one of the grave defects of their position.

It is worth mentioning then that if one represents the state of credal probability judgment characterizing the shared agreements of different individuals by the convex hull of the distributions in contention, judgments of independence are preserved.

In Levi (1978), I pointed out, just as Laddaga does, that for values of a greater than 0 and less than 1, $P_a(H/E)$ and $P_a(E)$ will differ even when H and E are independent according to P_1 and P_2 and $P_a = aP_1 + (1-a)P_2$. However, if one takes the set of values of $P_a(H/E)$ defined as $P_a(H \,\&\, E)/P_a(E)$ for all values of a and compares it with the values of $P_a(H/E)$ defined as $aP_1(H/E) + (1-a)P_2(H/E)$, it can be shown (as it was in Levi, 1978) that these two sets of values are identical. From this it easily follows that the set of values of $aP_1(H) + (1-a)P_2(H) = P_a(H)$ is the same set of values as the set of values of $P_a(H/E)$. Hence, a decision maker who finds out that E is true and conditionalizes will regard precisely the same set of probability assignments to H permissible after he has found out that E is true as he did before finding out. The discovery that E is true will be irrelevant to his deliberations as far as the truth of H and $\sim H$ are concerned.

Laddaga thinks that it is unnecessary to seek a consensus of shared agreements concerning probability judgment. He apparently has faith in the feasibility of identifying a pool of shared background knowledge or information relative to which one can justify an objectively grounded credal probability distribution following the approach of E. T. Jaynes. In my judgment, the discussion in Seidenfeld (1979) thoroughly undermines that approach. It is possible for rational agents to share all their information and yet honestly disagree in their credal probability judgments. Hence, so it seems to me, it is desirable to be able to formulate a consensus of shared agreements in probability judgments which can serve as the basis for inquiry.

To entertain the proposal I have made, one needs to allow for states of credal probability judgment which recognize more than one distribution to be permissible. Refusal to entertain this idea may have prevented Laddaga from appreciating the possibility of characterizing a conception of consensus as shared agreements which satisfies the demand that agreements concerning independence be preserved.

In any case, if we adopt a consensus as shared agreements on credal probabilities before the acquisition of evidence, we may hope to obtain new data via experimentation and observation which will yield a consensus which resolves the original dispute via inquiry. In typical cases, ample data will lead via Bayes's theorem and conditionalization to a reduction in the indeterminacy in the state of credal probability judgment. Consensus as the outcome of inquiry will be more determinate than the consensus as shared agreements adopted at the outset of inquiry (Levi, 1980, 13.7).

Thus, while disagreeing with Laddaga's trust in the possibility of finding prior credal distributions objectively grounded in the pool of evidence or information available to participants in a dispute before experimentation, I do agree with him that the right way to win consensus as resolution of dispute is through inquiry.

Collegiality and respect may be useful facilitators of inquiry where several individuals must participate; but they are no substitute for inquiry itself. Lehrer and Wagner appear to have forgotten that important truth.

REFERENCES

Berger, R. L.: 1981, 'A Necessary and Sufficient Condition for Reaching a Consensus by De Groot's Method', *Journal of the American Statistical Association* **76,** 415–418.

De Groot, M. H.: 1974, 'Reaching a Consensus', *Journal of the American Statistical Association* **69,** 118–212.

Hooker, C. A., J. J. Leach, and E. F. McClennen (eds.): 1978, *Foundations and Applications of Decision Theory,* Vol. 1, Reidel, Dordrecht.
Laddaga, R.: 1977, 'Lehrer and the Consensus Proposal', *Synthese* **36,** 473-477.
Lehrer, K.: 1978, 'Consensus and Comparison, a Theory of Social Rationality', in Hooker, Leach, and McClennen (eds.) (1978), pp. 283-309.
Lehrer, K. and Wagner, C.: 1981, *Consensus and Society,* Reidel, Dordrecht.
Levi, I.: 1974, 'On Indeterminate Probabilities', *Journal of Philosophy* **71,** 391-418. Reprinted with addendum in Hooker, Leach, and McClennen (eds.), (1978), pp. 233-261. Reprinted as Chapter 6 in this volume.
1977, 'Subjunctives, Dispositions and Chances', *Synthese* **34,** 423-455.
1978, 'Irrelevance', Hooker, Leach, and McClennen (eds.) (1978), pp. 263-275.
1980, *The Enterprise of Knowledge,* MIT Press, Cambridge, Mass.
1982, 'Ignorance, Probability and Rational Choice', *Synthese* **53,** 387-417.
Seidenfeld, T.: 1979, 'Why I Am Not an Objective Bayesian', *Theory and Decision* **11,** 413-440.

8

Compromising Bayesianism: a plea for indeterminacy

Jack Good reports in one of his papers that in 1947 he had a nonmonetary bet with M. S. Bartlett that the predominant philosophy of statistics one century later would be Bayesian. By 1979, Good modified his forecast. He then suggested that the predominant philosophy of statistics would be a Bayes–non-Bayes synthesis or compromise (Good, 1979). Presumably the basis both for the original forecast and its updated revision had less to do with extrapolations from trends then discernible in the thinking of his colleagues than with his optimistic assessment that Good thinking is the wave of the future together with his privileged relation to the source of this inspiration.

Good's predictions appear to be borne out to a considerable degree. Bayesianism has not silenced its critics among statisticians or philosophers who think about probability, induction and statistics. Still the contrast between the status of Bayesian ideas in the 1940s and now is astounding. More interesting yet, however, is the fact that even as the friends of Bayesianism multiply, there is also growing sympathy for ideas that are reminiscent of what Jack seems to have in mind by a Bayes–non-Bayes compromise.

Perhaps the major impetus for Good's quest for Bayes–non-Bayes compromises derives from the extremely demanding character of Bayesian ideals of rational belief, valuation and choice when it comes to applications. We cannot be expected to represent our probability assignments to hypotheses by precisely specified real numbers. The same applies to our utility judgments. Furthermore, Bayesian modeling of realistic situations often requires a complexity of structure which renders solutions to statistical decision problems difficult because of the costs of computation and, indeed, often because of the lack of computational techniques at

Reprinted from *Journal of Statistical Planning and Inference,* Vol. 25 (1990), Isaac Levi, "Compromising Bayesianism: A Plea for Indeterminacy," pp. 347–62, any price. Worshippers in the Bayesian temple must, perforce, compromise the principles of the true church when facing the real world.

Good was always sensitive to this point. Even when he was one of the lonely voices speaking out for a Bayesian point of view, he was thinking about ways and means to address questions about the application of Bayesian ideals. His discussions of dynamic probability and his deployment of the black box model in investigating upper and lower probability, type II rationality and minimax principles as supplements to the expected utility principle all reflect such concern and illuminate much of what he has in mind when he discusses Bayes–non-Bayes compromise. Indeed, Good is so open to exploring non-Bayesian considerations that the innocent reader may often wonder how good a Bayesian Good actually is.

I think that Good is a loyal Bayesian who has not compromised his Bayesian convictions with non-Bayesian principles at all. To explain why, I should like to present a view of probability judgment and decision making that does, indeed, compromise Bayesian ideals without rejecting them in toto and contrast it with Good's view. Superficially the formalisms which are associated with this alternative seem indiscernible from Good's own formalisms. I mean to show that both the formalisms and their intended applications differ in important ways.

Good's work was very important to me in developing my own ideas. I came to think about probability along the lines I now do by reflecting on the work of I. J. Good (1962), H. E. Kyburg (1961) and C. A. B. Smith (1961) on interval valued probability. Essentially I am reporting on the result of a dialectical process at the end of which I realized I disagreed rather substantially with someone whose writings were as responsible as any for initiating that very process. I hope Jack will take the report as a testimonial to my gratitude and not as another case of a dog biting the hand that fed him.

Good thinks that our judgments of comparative probability along with other data sometimes allow us to sort out inconsistencies in our probability judgments and to restrict assessments of numerical probabilities to those representable by functions belonging in some more or less clearly definable set. He suggests that identifying such sets of probability functions should be helpful in practical deliberation and inductive and statistical reasoning.

How are we to understand these sets of numerical probability functions? The situation is supposed to be analogous in important respects to the measurement of length or any other magnitude where, due both to the limitations of the measuring instruments available to us and to cost considerations, there are bounds on the precision with which the measurements can be carried out.[1] The length of some object may be measured to the nearest thousandth of an inch. One way of representing this is to report a range of values for the length. This representation may be constructed as meaning that the object has a numerically definite length in inches which is unknown but which falls within the interval specified. Some authors may be reluctant for metaphysical or epistemological reasons to posit the existence of a true but unknown length in inches but may, nonetheless, proceed as if there were one.[2] I do not care right now to enter into such metaphysical or epistemological disputes. My contention is that Good (and, I think, Koopman (1940) before him) understands interval valued belief or credal probabilities or families of probability functions as representing the result of an analogous measurement procedure.

The rational agent is supposed to have a 'black box' state of credal probability judgment representable by a single probability function or, at any rate, he should proceed as if he had such a probability function in his black box. The assumption is used, on the one hand, to eliminate inconsistencies in the judgments of comparative probability the agent makes and, on the other hand, is taken for granted when efforts are made to elicit the properties of this unknown probability judgment: The limitations of our measurement procedures and the costs of carrying out the elicitations impose severe constraints on the precision with which the agent himself or some observer can measure his probabilistic degrees of belief. Credal probabilities will be rounded off after a few decimal places even under the best of circumstances. The most one can hope for is a set of probability functions each of which is a conjecture as to the probability function which represents the agent's unknown credal state.

Good's interest in this problem is entirely sound. It is painfully clear that, as flesh and blood, we face overwhelming problems in satisfying demands of rationality. It is not to be expected that we know our belief states in full precision or identify those beliefs we ought, as ideally rational agents, to have. This does not mean, however, that we can dispense with notions of ideal rationality or states of probability judgment which satisfy them. Good's practice of appealing to a conception

1 A relatively recent and explicit statement of this view is found in Lindley, Tversky and Brown (1979), pp. 146–147.

2 See, for example, the note by J. M. P. Moss in the discussion of Lindley, Tversky and Brown (1979), p. 176, and Lindley's reply on p. 178.

of an ideal belief state in the black box meeting full standards of first order rationality (to which the agent is committed even though incapable of fully living up to the commitments undertaken) is required in order to give coherence to the study of second order rationality. Without first order rationality, second order rationality is blind just as without second order rationality, first order rationality is empty.

What is open to discussion is the specification of the contents of the black box which characterizes the conception of first order rationality. Good's view presupposes that we and the agent should proceed as if the agent is committed to a belief state represented by a single probability distribution but that the agent (or the observers of the agent's behavior) has only partial knowledge of that state of probabilistic judgment.

According to Good's idea, the judgments of comparative probability which an agent can make, when they are consistent, define a nonempty set of probability functions eligible for consideration as strict Bayesian black box representations of his state of credal probability judgment.

I too mean to use a set of probability functions to characterize the agent's credal state. Instead, however, of looking at functions in the set as possibly true hypotheses about the unknown strictly Bayesian credal state, I suggest that the set consists of distributions all of which are permissible to use in assessing risks. Even if the agent were perfectly self aware, he might deny that exactly one probability function is permissible. The agent is indeed in a state of probabilistic ignorance; but the ignorance does not concern which of rival conjectures concerning the agent's strictly Bayesian credal state is true. The agent is clear that he or she is not in a strictly Bayesian credal state representable by a unique probability function. The agent's ignorance consists in the recognition of the permissibility of more than one probability function in the evaluation of policies in practical deliberation and scientific inquiry. In such a state of credal ignorance, the set of permissible probability functions is not a reflection of an imprecision in the measurement of numerically determinate credal probabilities but represents a numerical indeterminacy in the agent's probability judgments themselves. (See Levi, 1974, footnote 7.)

In real life, agents who are not strict Bayesians because their credal states are numerically indeterminate will fail to be clear about the identity of the sets of distributions which represent their credal states. There will be imprecision in the identification of indeterminate probabilities. In this respect, there is no difference between the outlook of those who insist on the rationality of indeterminate probability judgment and strict Bayesians who deny it. They both agree that the specification of credal states will perforce be imprecise. The serious issue under dispute is whether one should acknowledge the reasonableness of states of indeterminate probability judgment as I wish to do or take the strict Bayesian approach and prohibit such credal states.

Once Good's point that judgments of probability are imprecise is granted, what is the importance of worrying as to whether credal states would be indeterminate were imprecision in measurement eliminated? If anything is an idle counterfactual question, this looks like it.

Appearances, however, are deceiving. The strict Bayesian view and the 'quasi Bayesian' view which allows for indeterminacy have different implications for decision making and for inquiry. The differences will be explained in two different ways. I shall appeal to an analogy to issues arising in group decision making, on the one hand, and to ramifications for criteria for individual decision making on the other.

When two or more agents disagree concerning what is true, concerning how probable it is that a hypothesis is true or how desirable that it be true, they may be prepared to remain in disagreement and treat each other's views with contemptuous toleration. The mere fact that someone disagrees with one's judgments is insufficient grounds for opening up one's mind. Most epistemologists forget that determining when we are justified in opening up our minds to partially baked ideas is as urgent a question as deciding when we are justified in closing our minds once opened. Perhaps we ought to tolerate dissent from our convictions no matter how hare-brained that dissent might be. Still there is an important difference between the contemptuous toleration we accord opinions that are nonstarters and those conjectures that merit our respect and provide the occasion for serious reflection and inquiry.

On the assumption that dissenting opinions sometimes (though not always) merit serious respect, the participants to the dispute sometimes (though not always) are justified in initiating some sort of investigation to settle their dispute. An early step in such a joint effort is the identification of those shared agreements which might serve as the noncontroversial basis for subsequent inquiry. Once this is done, investigation may proceed relative to that background of shared agreements according to those methods acknowledged to be appropriate to the problem under consideration.

On this view of disagreement and its resolution through inquiry, two distinct senses of consensus can be identified:

a. One may speak of that consensus of the participants at the beginning of inquiry that constitutes the background of shared agreements on which the investigation is initially grounded.

b. At the termination of inquiry, the parties to the investigation reach a consensus. This consensus may indicate the views of one of the parties to the dispute prior to inquiry or run counter to the initial views of all of them. What is clear is that if inquiry is successful, this consensus as the outcome of inquiry will be quite different from the consensus as shared agreement with which a properly conducted inquiry should begin.

Talk of consensus is appropriate when we are thinking of groups of individuals who may disagree with one another. However, when an individual agent is in doubt as to the merits of conflicting viewpoints, his predicament is analogous in relevant respects to the situation of a group. At the outset of inquiry, he should rely only on those assumptions which are not in dispute between the conflicting viewpoints and at the end of inquiry it is to be hoped that he will attain a resolution of his doubts. This contrast between the settled assumptions adopted at the onset of inquiry and the resolution of the issue at its termination corresponds well to the contrast between two types of consensus in group deliberation. The analogy should be kept in mind during the course of this discussion.

The contrast between consensus as shared agreement at the onset of inquiry and consensus as the outcome of inquiry will seem unimportant to someone who denies the feasibility of conducting inquiry into certain types of problems without begging the issues under dispute. As I understand him, T. Kuhn (1962) thinks that controversies over competing 'paradigms' are of this kind. To shift to a system of shared assumptions and investigate the merits of rival hypotheses is not feasible. The system of shared assumptions, if intelligible at all, is far too weak to form a basis for subsequent inquiry aimed at deciding between rival claims.

From the Kuhnian point of view, there is only one kind of consensus that is of interest: the consensus reached when the revolution is over and one view has emerged as the victor to guide inquiry in the subsequent period of normal science. This is consensus neither as shared agreement at the onset of inquiry nor as outcome of inquiry. It is a surrogate for the latter in that it is a settling of the dispute. But in this case, the dispute is settled by revolution, conversion, gestalt switch or, perhaps, by some other psychological, social or political process. The outcome is not a product of genuine inquiry in which pros and cons are weighed from a point of view which begs no questions under scrutiny.

I shall take for granted without further discussion that the Kuhnian outlook should be rejected, that it is desirable wherever feasible to resolve disputes by engaging in inquiry based on shared agreements that beg no controversial issues. When this approach is taken, the difference between consensus as shared agreement at the onset of inquiry and as outcome of inquiry is relevant, important and clear.

This attitude ought not to be restricted to controversies concerning what is to be taken for granted and, hence, assigned probability 1. If two or more physicians are responsible for the care of a patient, they may share the same evidence and agree on the correct diagnosis of the disease offered. They could, nonetheless, disagree in their credal probabilities for the success of therapies I, II and III effecting a cure. Such disagreement may be severe enough for them to recommend treating the patient in different ways. In such a case, where the physicians must act in concert, it may be desirable that they act on the basis of a consensual probability judgment.

If the physicians had the opportunity to collect data before coming to a decision and if they could reach some sort of initial consensus with respect to their probabilities, they could combine the 'prior' consensual judgment with the data and update via Bayes' theorem to reach a final judgment which they could then deploy in the treatment of the patient.

The initial consensus, so I maintain, should be a consensus as shared agreement whereas the final judgment would be a consensus reached as the outcome of inquiry. The conceptual problem to be faced is to characterize these two types of consensus in probability judgment.

In the real world, the physicians will often have to reach a decision about treating the patient without the benefit of a prior inquiry. I contend that in the absence of grounds for resolving their dispute, they should beg no questions and should base their decision on a consensus probability judgment in the sense of consensus as shared agreement. This consideration furnishes an additional reason for interest in characterizing consensus as shared agreement.

Suppose the options were lotteries rather than therapies and that there are two individuals i and j in place of the physicians who must decide together which of the options to choose. They are told that a lottery is to be conducted where the chance of event E occurring is either 0.9 or 0.1. The three options are as follows:

Case A. *Option I:* Receive 55 utiles if E occurs and lose 45 if it does not.
Option II: Lose 46 utiles if E occurs and win 55 if it does not.
Option III: Receive 0 utiles for sure.

Individual *i* has a strictly Bayesian credal state for *E* and ~*E*. He assigns credal probability of 0.9 to the hypothesis $H_{0.9}$ that the chance of *E* occurring is 0.9 so that his credal probability for *E* is 0.82. The expected utility of option I is 37, of option II is −27.36 and III, of course, 0.

Individual *j* assigns credal probability 0.1 to $H_{0.9}$ so that his credal probability for E is 0.18. The expected utility of option I is −27, of option II is 37.82 and of option III is 0.

Left on his own, *i* would choose I. *j* would choose II. But they must choose in concert just as the physicians must.

Here, too, a contrast between consensus as shared agreement and as outcome of inquiry is apparent. If the two parties have the opportunity to experiment by running independent trials to test whether the chance of *E* is 0.9 or 0.1, they must invoke some consensus on prior probabilities; this will, perforce, be a consensus as shared agreement.

I propose to represent the consensus as shared agreement of *i* and *j* not by a single distribution but by a set of probability assignments to the rival hypotheses $H_{0.9}$ and $H_{0.1}$ about the chances of *E*. This set of permissible distributions should consist of all the weighted averages of the assignments made by *i* and *j*. That is to say, the set of probability functions should be the convex hull of the distributions representing the opinions of *i* and *j* or, more generally, the individuals participating in the consensus.

Why should the convexity condition be imposed? The first thing to keep in mind is that credal probabilities are used to evaluate feasible options with respect to expected utilities. When more than one probability function is permissible, more than one ranking of the options with respect to expected utility will normally be permissible. I assume that no option ought to be chosen unless it has a maximum expected utility relative to at least one ranking with respect to expected utility. That is to say, no option is admissible unless it is admissible with respect to expected utility (*E*-admissible).

Individuals *i* and *j* seek a consensus as shared agreement in probability judgment in order to obtain a shared agreement in their evaluation of the feasible options with respect to expected utility. In the case under consideration, the utilities assigned payoffs for the two lotteries are the same for both *i* and *j*.

Given this common utility function v and the two credal probability functions p_i and p_j, the expected utility functions ev_i and ev_j are determined for all mixtures of options I, II and III. The conception of consensus suggests the following conditions on permissible expected utility functions.

- *Potential resolution:* (a) ev_i and ev_j are both permissible expected utility functions. (b) Every potential resolution of a conflict between the rankings determined by ev_i and ev_j is permissible.
- *Bayesianism:* Every potential resolution of the conflict between ev_i and ev_j is an expected utility function determined by the common utility function v and the probabilities p and p_j respectively.
- *Pareto unanimity:* If x and y are mixtures of options I, II and III and $ev_i(x) \leq ev_j(y)$ and $ev_j(x) \leq ev_j(y)$, then every expected utility function ev which is a potential resolution of the conflict between these two functions satisfies the condition that $ev(x) \leq ev(y)$.

The Bayesian requirement entails that the expected utility functions over the set of mixtures of I, II and III which are potential resolutions of the conflict are von Neumann–Morgenstern utilities.

Because all such potential resolutions obey Pareto unanimity, by adapting arguments of M. Fleming (1952) and J. Harsanyi (1955), we can show that every potential resolution ev must be a weighted sum of ev_i and ev_j.

Moreover, every such weighted sum will be an expected utility determined by v and a weighted sum of p_i and p_j. Hence, every weighted sum satisfies both Bayesianism and Pareto unanimity and qualifies as a potential resolution. Hence, Potential resolution implies that the set of permissible expected utility functions should be the convex hull of the expected utility functions i and j.[3]

3 If i and j share a common probability but different utilities, the Bayesianism condition can be modified in the obvious way and an argument parallel to the one sketched above may be produced to show that the set of permissible utility functions in consensus as shared agreement should be convex.

Problems arise when i and j disagree in both their utility and probability functions. Subsequent to my delivering this paper, Seidenfeld, Kadane and Schervish obtained an important and surprising result for this case. Consider the set of mixtures of 'horse lotteries' in the sense of Anscombe and Aumann (1963) generated by the prizes and states figuring in options I, II and III (so that these options and all mixtures of them are in the set) and replace the Bayesianism requirement by the condition that the Anscombe-Aumann axioms are satisfied by all rankings with respect to expected utility which are potential resolutions of conflict. Anscombe and Aumann showed that there is representation of any ranking meeting their axioms by an expected utility function determined by a probability p for the 'states' and state independent utility v where the probability is unique and the utility is unique up to a positive affine transformation. If we assume that the parties i and j to the consensus both have state independent utilities and suppose that every potential resolution should be representable, in that case, by a probability-utility pair where the utility is state independent and, if in addition we require Pareto unanimity, Seidenfeld, Kadane and Schervish (1989) show that only the expected utility functions of i and j are permissible.

Seidenfeld, Kadane and Schervish claim that in consensual decision making Pareto unanimity concerning evaluations of available options is sacrosanct as is the Bayesianism requirement on permissible rankings of options. They conclude, therefore, that

Since we are focusing on cases where the utilities for the payoffs on consequences are the same for all participants in the decision, the convexity of the set of expected utility functions entails the convexity of the set of probability functions used to compute the expectations.

In case A, options I and II are *E*-admissible. Option III is not. There is no weighted average of the expected utility functions for *i* and *j* which ranks option III as best.

Contrast this predicament with the following variation:

Case B. *Option I**: Receive 45 if *E* occurs. Lose 55 if not.
*Option II**: Lose 54 if *E* occurs. Receive 45 if not.
*Option III**: 0 for sure.

According to *i,* the expected utility of I* is 27, of II* is −36.18 and of III is 0. According to *j,* the expected utility of I* is −36.45, of II* is 27.18 and of III* is 0.

I* and II* are *E*-admissible. But so is III*. This is so because III* bears higher expected utility than the other options when the probability for

when *i* and *j* differ both in probability and utility judgement, convexity of the set of permissible expected utility functions, the set of probability functions and the set of utility functions should be abandoned.

In my judgment, their view is incompatible with certain fundamental requirements on consensus as shared agreement. Suppose *i* and *j* differ both in probability and utility judgment. The disagreement, however, derives from the fact that *i* is certain that *e* is true whereas *j* is certain that *e* is false. Both *i* and *j* agree, however, as to what their conditional credal probabilities for $H_{0.9}$ given *e* and given ~*e* would be were they both to move to a position of suspense as to whether *e* is true or false. Moreover, both *i* and *j* agree as to what probability to assign to *e* in both a state of suspense.

I contend that in such a situation, if *i* and *j* are to move to consensus as shared agreement, they should move to a position of suspense between *e* and ~*e*. However, it is demonstrable, by the very arguments advanced by Seidenfeld, Kadane and Schervish, that there are pairs of options in the space of horse lotteries which *i* and *j* initially ranked the same way but which in the consensus have their order reversed.

The moral of the story seems to be that Pareto unanimity needs to be modified. I suggest that we require a Robust Pareto unanimity. Such unanimity is robust for those cases of agreement of preference between two options which is preserved when *i's* probability is combined with *j's* utility and vice versa. Robust Pareto unanimity is obeyed even in the case where *i* and *j* move to suspense between *e* and ~*e*. If we require Bayesianity and Robust Pareto unanimity, it can be shown that the convexity of the set of permissible probability functions and the set of permissible utility functions can be secured. In the special case considered in the text where there is a shared utility but different probabilities, Robust Pareto unanimity and Pareto unanimity are equivalent, as they are when the probabilities are shared and there is disagreement in utility.

I intend to elaborate on my disagreement with Seidenfeld, Kadane and Schervish elsewhere but I wish to emphasize my admiration for the fundamental contribution they have made to the discussion of consensus and group choice.

$H_{0.9}$ is 0.5. In the first scenario, option III never bears maximum expected utility.

If the set of permissible probability distributions consisted exclusively of the values for $x = 0.9$ and 0.1 without the intermediate values, there could be no difference between the two scenarios. Yet, there does appear to be an important difference in the two cases.

These examples illustrate the force and value of requiring a convexity condition on sets of permissible credal probability functions representing consensus in probability as shared agreement.

Consensus as the outcome of inquiry presents us with a different picture. Given that *i* and *j* have a disagreement in their probability judgments, they might hope at the end of inquiry to reach an agreement on a single probability distribution or, if not that, they might obtain a convex set of distributions whose elements are not very different from one another.

Thus, *i* and *j* might be able to update the initial consensus as shared agreement by Bayes theorem and conditionalization to obtain a convex set of posterior distributions which is more determinate than the family of prior probabilities in the initial consensus. Under appropriate conditions, the set of posterior distributions converge on a unique distribution – indeed to unanimous assignment of probability 1 to either $H_{0.9}$ or $H_{0.1}$. (For a recent exploration of the 'appropriate conditions', see Schervish and Seidenfeld (1987).)

Thus far, I have been speaking of situations where agents are concerned to act in concert. I don't claim to be talking about all forms of group decision making but only about what may be called 'liberal group decision making'. In such cases, agents concerned to act in concert begin by identifying their shared agreements and then undertaking efforts to resolve their disagreements through non-question begging inquiry from shared agreements.

There is a vision of individual decision making which parallels this model of liberal group decision making. Just as it is acceptable for groups to be committed to indeterminate states of credal probability judgment representable by nonempty convex sets of probability functions which the consensus recognizes as permissible, the same should be entertainable for individual agents. A physician may after all be in doubt as to which of two probability-of-success distributions to assign to therapies *A*, *B* and *C* and, pending resolution of his conflict, may recognize all distributions which are weighted averages of these two distributions to be permissible. By adopting such an indeterminate credal state, the agent is endorsing what is, in effect, shared in common by the rival distributions with respect to how such distributions are to be used in decision making.

The thesis I have been leading up to is this: Counter to the strict Bayesian view, the probabilistic beliefs of neither individual nor group decision makers need be representable by a uniquely permissible probability function. Rational agents not only do not lose their rationality but actually display it by being in that state of doubt where many probability distributions are recognized as permissible.

Strict Bayesians are committed to denying the rationality of endorsing indeterminate states of credal probability judgment. As a consequence, they are precluded from considering liberal group decision making. There can be no consensus as shared agreement at the onset of inquiry and, hence, no consensus as the outcome of inquiry.[4]

Strict Bayesians can and do countenance consensus as the outcome of revolution, conversion, bargaining or some other psychological, social or formal process of conflict resolution. Consensus in one of these senses

4 One approach to the problem of ascertaining unknown credal probabilities is to treat the agent as an experimental subject whose overt probability judgments are data relevant to the unknown credal state. The observer or investigator is to approach the task of estimating the agent's true credal probability state as a good Bayesian would. Lindley, Tversky and Brown (1979) adopt this view. Their 'external' approach addresses what French (1985) calls the 'expert problem' and their basic method resembles the approach French subsequently follows. A decision maker consults a panel of experts for their probability assessments of a certain set of events before reaching a decision. The idea is to regard the testimony of the experts as data to be fed into a Bayesian calculation.

Whether the testimony of witnesses or experts should be used as *evidence* for a hypothesis as these authors suggest or as *input* into a program for routine decision making (Levi, 1980, ch. 17) is a delicate question beyond the scope of this discussion to consider.

However, it is important to distinguish, as French clearly does, between the expert problem and the group problem where individuals of differing opinions wish to make a joint decision. French examines the idea of introducing a 'supra decision maker' who is seeking to make policy in terms of the interests of the members of the group. The supra decision maker uses the attitudes of the members as data to update his own prior beliefs over the relevant hypotheses. His problem is, in effect, reduced to a version of the expert problem. As French rightly observes, however, the supra decision maker does not exist. French ends his discussion of group decision making questioning whether there is any sense at all to the notion of group decision making.

The notion of a supra decision maker illustrates one way in which a question concerning consensus as shared agreement can be reconstructed as something resembling a measurement problem. For that is exactly the so-called expert problem as understood by French and by Lindley, Tversky and Brown. French seeks to reduce consensus as shared agreement in this way, doubts that it can work and concludes that the intelligibility of the notion of group decision making is doubtful. Surely this is question begging. Why not conclude, as I would, that strict Bayesian requirements of rationality need modifications?

may be the outcome of a process of conflict resolution. But it is not the outcome of that specific type of process I am calling non-question begging inquiry.

Why are strict Bayesians prevented from allowing for liberal inquiry? Because they cannot recognize the legitimacy of consensus as shared agreement. For a strict Bayesian, such a consensus must be represented by a uniquely permissible probability function. But when two or more individuals have differing probabilistic opinions, a single probability distribution cannot represent a consensus as shared agreement.

To appreciate why a single probability distribution cannot represent consensus as shared agreement, we have to recall the shared agreement it is desirable to preserve.

We have already taken note of one such type of shared agreement. Two agents who differ in their probabilities but share the same utility function over the outcomes of alternative policy choices will have different expected utilities. In that setting, we should require shared agreements in probabilities to mirror shared agreements in expected utility. These shared agreements are the agreements in the way available options and mixtures of these are ranked with respect to expected utility. That is to say, Pareto unanimity ought to be obeyed. But by the argument of Harsanyi previously mentioned, the only function of utilities which is itself a utility and obeys the Pareto unanimity condition is a weighted average of utilities. This entails that if the consensus probability as shared agreement is to capture the agreements in expected utility, it must be a weighted average as the so-called linear opinion pool conception requires.

This, in itself, does not imply that a strict Bayesian cannot represent consensus as shared agreement. Any weighted average of the opinions of parties to consensus is going to be a strictly Bayesian probability.

However, shared agreements in expected utility are not the only shared agreements which ought to be preserved in consensus as shared agreement. For example, if two or more agents regard events X and Y as probabilistically independent, the state of probability judgment representing consensus as shared agreement ought to acknowledge this. A weighted average or linear opinion pool cannot.

Logarithmic opinion pools and variants on them have been proposed which can preserve shared agreements concerning independence as well as such features as 'external Bayesianity'. But whatever other merits such proposals may have, they cannot represent consensus as shared agreement; for Pareto unanimity on expected utility will not be obeyed.

The upshot is that no single probability distribution can represent consensus as shared agreement. On the other hand, the convex hull of the probabilistic views of all parties to a dispute satisfies the Paretian requirement, preserves agreements on probabilistic independence (Levi, 1978 and 1980, pp. 225–233) and satisfies cognate conditions like external Bayesianity. Once one allows for probabilistic indeterminacy along the lines I have been suggesting, liberal group decision making beginning with consensus as shared agreement becomes coherent. And individual decision making with probabilistic suspense or ignorance is likewise made intelligible.

Thus, strict Bayesians are obliged to deny the properties of consensus as shared agreement. If one insists that the group decision making ought to proceed according to strict Bayesian rules, consensus as shared agreement is prohibited. Similarly if one insists that rational individuals ought to adopt states of credal probability meeting the strict Bayesian requirement that exactly one probability function is permissible to use in assessing risks and expectations, rational individuals are prohibited from being in doubt or suspense between rival opinions.

As we have seen, even strict Bayesians do acknowledge a certain form of ignorance about personal probabilities. It is the kind of ignorance to which Good addressed himself when he focused attention on the imprecision in our assessments of probability due to lack of self knowledge, computational incapacity, limitations of memory and costs of deliberation.

In the group decision problem, it should be obvious that the convex set of distributions representing the consensus as shared agreement between *i* and *j* is not the product of anything like imprecision in measurement. To say otherwise, we would have to proceed as if among the probability functions there were exactly one which represented the consensus. As we have seen, no single function can represent a consensus as shared agreement among other probability functions. Hence, the indeterminacy involved in consensus as shared agreement cannot be reduced to imprecision in the measurement of a strictly Bayesian credal state.

It may, perhaps, be thought that indeterminacy in probability judgment is acceptable in group decision making but is to be denied rational status in individual decision making. I fail to understand what relevant difference there is between the two.

In the light of the distinction between imprecision and indeterminacy in credal probability judgment I have been pressing, how are we to understand the notion of a Bayes-non-Bayes compromise?

Consider then the first of the two choices between lotteries I describe – i.e., case A. Options I and II are both *E*-admissible. Option III is not. If, however, we compare I and II with respect to the minimum risk incurred (in the sense of Wald (1950)), option I is superior to option II and might, for that reason, be chosen.

This is a form of Bayes–non-Bayes compromise. We first attempt to identify those options which are *E*-admissible. If there are several of these, considerations of expected utility cannot decide between them. In such a case, a non-Bayesian decision criterion such as minimax risk may be invoked.

The version of minimax risk deployed as a secondary criterion differs from other forms of Bayes–non-Bayes compromise in at least two important respects. The minimax criterion is not applied to all the options but only the *E*-admissible ones. Otherwise option II would be chosen as Gärdenfors and Sahlin (1982) recommend. Although acknowledging indeterminacy deviates from the requirements of strict Bayesianism, I do wish to preserve one strand of Bayesian thinking: Admissible options maximize expected utility according to some permissible ranking with respect to expected utility. Maximizing minimum expectation as Gärdenfors and Sahlin propose fails to do this (unless one unrealistically includes all mixtures of available options among the available options).

Another kind of Bayes–non-Bayes compromise is proposed by Good. In discussing Wald's minimax risk criterion, Good also suggested that we might restrict its application to options which are Bayes solutions – i.e., come out best in expected utility according to some function in the set (Good, 1952). However, Good understands this proposal differently than I do. According to Good, invoking Wald's criterion is one way (not the only one) of selecting a probability for $H_{0.9}$ to use as one's uniquely permissible credal probability for the purpose of determining expectations. As Wald himself suggested, minimax is a pessimistic strategy and that is revealed by the choice of a least favorable distribution to use in reaching a decision.

I do not understand the matter in this way and this explains how my Bayes–non-Bayes compromise differs from Good's. The use of the minimax criterion as I conceive of it does not imply that the agent is proceeding as if his personal probability were the least favorable distribution. Minimax is a criterion used to decide between options when one's credal state is indeterminate and considerations of expected utility are of no further help. One is not in any sense committed to choosing as if any specific distribution is uniquely permissible. Only someone who used sets of distributions to represent imprecision in probability judgment could think so.

Suppose that in case A, option II is withdrawn before the choice of option I is implemented and the choice is reconsidered. Both options I and III are then *E*-admissible and considerations of security favor choosing III. But the probability distributions against which III maximizes expected utility vis-à-vis I are different from those against which I maximizes expected utility vis-à-vis II and III. The agent, however, has not modified his credal state due to changes in the set of available options. Both the distributions assigning maximum expected utility to II in the pairwise choice and the distributions assigning maximum expected utility to I in the three way choice are permissible in both contexts of choice.

The 'choice function' generated by applying minimax to *E*-admissible options violates the so-called independence of irrelevant alternatives, or property α as Sen (1970) calls it (see Levi, 1974). Property α asserts that if option x is a member of feasible set S which is a subset of set T, and x is a member of the set of admissible options $C(T)$ in T, x is a member of $C(S)$. Given that the criteria of admissibility violate α, the revealed preference relation which is generated (see Levi, 1986b) lacks the properties crucial to furnishing a weak ordering of options. That should not be surprising since the recognition of indeterminacy in probability should imply that options are not weakly ordered with respect to expected utility.

For Good, however, the indefiniteness in our probability judgments merely reflects a lack of precision in our measurement of probability – i.e., our incapacity to identify the ordering which is in the black box. Although he concedes that we may use non-Bayesian considerations such as appeal to minimax so that we may proceed as if we had precision in our probability and utility judgments, he would object to the recommendation that I be chosen in the three way choice and II chosen in the two way choice. Once one has used minimax to choose I in the three way choice, one has committed oneself to proceeding as if the probability of $H_{0.9}$ is greater than 0.495. This commitment precludes choosing option III in the pairwise choice.

Thus, for Good, a Bayes–non-Bayes compromise is not intended to compromise the ideals of strict Bayesianism except in those cases where it is beyond human capacity to avoid doing so (which is not the case in the example under discussion). Recognizing indeterminacy in probabilities (and utilities) allows for more far reaching compromise of Bayesianism. The ideals themselves are compromised.

If I am right in my interpretation of Good's view, he appears to be a loyal strict Bayesian after all – at least with respect to individual choice.

I want to say that both in individual choice and liberal group choice, the same standards of rationality ought to prevail. Indeterminacy in credal probability in the individual case corresponds to consensus as shared agreement in the group case. If one thinks, as I do, that consensus as shared agreement is often a desideratus of group rationality, it should sometimes be a desideratum of individual rationality as well.

On this view, it becomes possible to rationalize types of behavior which Bayesians must perforce find deviant even though men who are prima facie sane, intelligent and informed about their predicaments persist in the deviations. This is true for the Allais and Ellsberg paradoxes, for the other so-called preference reversal phenomena and for Newcomb's paradox (Levi, 1986a, 1986b).

Once indeterminacy in credal probability is recognized, we may also begin to bridge some of the conceptual gaps which separate Bayesian from non-Bayesian approaches to probabilistic and statistical inference. (This is a leading theme of Levi, 1980.) When great anti-Bayesians like Fisher, Neyman and Wald denied the existence of prior probabilities, indeterminists may reinterpret these authors as recognizing all probability distributions over the relevant space of hypotheses (parametrized or not) as being permissible so that the prior credal state is maximally indeterminate.

Perhaps these authors were mistaken in rigidly insisting on such maximal indeterminacy. I myself think they were (Levi, 1980). But that is no reason for endorsing the opposite rigidity and insisting that priors should be maximally determinate and that the only obstacle to numerical definiteness is imprecision in measurement. A less dogmatic intermediate approach recognizes that priors might be partially indeterminate in varying degrees. When that is the case, it remains open to us to reduce indeterminacy through the collection of data and updating by Bayes theorem. Beginning with consensus as shared agreement, we may reach a fairly definite consensus as outcome of inquiry.

One of the attractions of strict Bayesianism is its great generality and the promise it affords of systematizing so much of our thinking about rational inference and choice. My contention is that if one recognizes indeterminacy in probability judgment and the decision criteria appropriate to it, we can preserve the attractive features of strict Bayesianism if for no other reason than that the strict Bayesian theory becomes a special case of a more general point of view. At the same time, we can recognize as rational, forms of behavior which honest and intelligent decision makers wish to insist are acceptable but which strict Bayesians must condemn as irrational.

None of this, of course, detracts from the importance of Good's contributions to our understanding of imprecision in probability judgment. Nor do I want to be guilty of reducing imprecision in probability judgment to a species of indeterminacy. Imprecision and indeterminacy are two distinct forms of indefiniteness displayed in probability judgment and present in deliberation and inquiry. My aim here has been to make a plea for serious consideration of indeterminacy in probability judgment as distinct from imprecision and for further consideration of the ramifications of doing so.

REFERENCES

Anscombe, F. J. and R. J. Aumann (1963). A definition of subjective probability. *Ann. Math. Statist.* **34,** 199-205.

Fleming, M. (1952). A cardinal concept of welfare. *Quar. J. of Econ.* **66,** 366-884.

Gärdenfors, P. and N.-E. Sahlin (1982). Unreliable probabilities, risk bearing and decision making. *Synthese* **33,** 361-386 (reprinted in Gärdenfors and Sahlin, 1988).

(Eds.) (1988). *Decision, Probability and Utility.* Cambridge University Press, Cambridge.

Genest, C. and J. B. Kadane (1984). Combination of subjective opinion, an application and its relation to general theory. Tech. Report No. 329, Dept. of Statistics, Carnegie Mellon University.

Genest, C. and J. V. Zidek (1984). Combining probability distributions. A critique and annotated bibliography. Tech. Report No. 316, Dept. of Statistics, Carnegie Mellon University.

Good, I. J. (1952). Rational decisions. *J. Roy. Statist. Soc. Ser. B* **14,** 107-114.

(1962). Subjective probability as a measure of a non-measurable set. In: E. Nagel, P. Suppes and A. Tarski (Eds.), *Logic, Methodology and Philosophy of Science.* Stanford University Press, Stanford, CA, 319-329 (reprinted in Good, 1983).

(1979). Some history of the hierarchical Bayesian methodology. In: J. M. Bernardo et al. (Eds.), *Bayesian Statistics.* University Press, Valencia, 489-519 (reprinted in Good, 1983).

(1983). *Good Thinking.* University of Minnesota Press, Minneapolis, MN.

Harsanyi, J. (1955). Cardinal welfare, individualistic ethics, and interpersonal comparisons of utility. *J. of Pol. Econ.* **63,** 309-321.

Koopman, B. O. (1940). The base of probability. *Bull. Amer. Math. Soc.* **46,** 763-774.

Kuhn, T. (1962). *The Structure of Scientific Revolutions.* University of Chicago Press, Chicago, IL.

Kyburg, H. E. (1961). *Probability and the Logic of Rational Belief.* Wesleyan University Press, Middletown, CT.

Levi, I. (1974). On indeterminate probabilities. *J. of Philos.* **71,** 391-418 (reprinted in Gärdenfors and Sahlin, 1988, and as Chap. 6 of this volume).

(1978). Irrelevance. In: C. A. Hooker, J. J. Leach and E. F. McClennen (Eds.), *Foundations and Applications of Decision Theory* Vol. 1. Reidel, Dordrecht-Boston.

(1980). *The Enterprise of Knowledge.* MIT Press, Cambridge, MA.

(1986a). The paradoxes of Allais and Ellsberg. *Econ. and Philos.* **2,** 233–263 (reprinted as Chap. 10 of this volume).

(1986b). *Hard Choices.* Cambridge University Press, Cambridge.

Lindley, D. V., A. Tversky and R. V. Brown (1979). On the reconciliation of probability assessments (with discussion). *J. Roy. Statist. Coc. Ser. A* **142,** 146–180.

Schervish, M. J. and T. Seidenfeld (1987). An approach to consensus and certainty with increasing evidence. Tech. Report N. 389, Dept. of Statistics, Carnegie Mellon University.

Seidenfeld, T., J. B. Kadane and Schervish, M. J. (1989). On the shared preferences of two Bayesian decision makers. *J. Philos.* **86,** 225–244.

Sen, A. K. (1970). *Collective Choice and Social Welfare.* Holden Day, San Francisco.

Smith, C. A. B. (1961). Consistency in statistical inference and decision (with discussion). *J. Roy. Stat. Soc. Ser B* **23,** 1-25.

Wald, A. (1950). *Statistical Decision Functions.* Wiley, New York, pp. 16–31.

9
Pareto unanimity and consensus

T. Seidenfeld, J. B. Kadane, and M. Schervish[1] have published an important paper on consensual decision making. These authors contend that the consensual preference among available options ought to preserve those comparisons among available options which all parties to consensus agree in making. Pareto unanimity ought, in this sense, to be a constraint on consensual decision making.[2]

Seidenfeld, Kadane, and Schervish then consider situations where there are two agents (they speak of Dick and Jane) concerned to fix on a consensus, who face so-called horse lotteries in the technical sense of F. J. Anscombe and R. Aumann[3] over two states. If Dick and Jane share

Thanks are due to Teddy Seidenfeld for important error detection.

From *The Journal of Philosophy* LXXXVII, 9 (September 1990), pp. 481-92. Reprinted by permission of The Journal of Philosophy.

1 "On the Shared Preferences of Two Bayesian Decision Makers," *The Journal of Philosophy,* LXXXVI, 5 (May 1989): 225-244.

2 Consensual decision making is a species of group decision making where a choice among a set of options is based on the shared agreements among the parties to consensus concerning (i) what is taken for granted as settled or certain, (ii) probability judgments, and (iii) values or utilities. Consensual decision making ought to satisfy minimal conditions of rationality imposed on individual decision makers. This requirement need not hold for other forms of group choice where each member of the group makes a decision according to his respective beliefs and desires and the "group" choice is the product of these individual choices.

Seidenfeld, Kadane, and Schervish agree with the view I favor that, if the parties to consensus order the available options according to different preference rankings over all horse lotteries generated by a given set of states and prizes (see footnote 3), not only should the consensus respect their shared agreements, it should not impose a preference over the horse lotteries where the parties to consensus disagree. The consensus should be represented by all weak orderings that preserve the shared agreements. See footnote 4.

3 "A Definition of Subjective Probability," *Annals of Mathematical Statistics,* XLIX (1963): 199-205. A roulette lottery over a set of prizes is a stochastic process which when implemented yields prizes from the given set with specific objective or statistical probabilities or chances. A horse lottery relative to a specified list of states and prizes is an option which, for each state, yields a roulette lottery on the prizes as the payoff. Each horse lottery can be represented as a function from the states to the roulette lotteries. Given a weak preference ordering over all horse lotteries relative to a finite set of states and a finite set of prizes satisfying von Neumann-Morgenstern conditions

the same probability distribution over the states but different utility assignments to the prizes, Pareto unanimity requires that every utility function defined over the prizes which is a permissible resolution of the differences between Dick's utility and Jane's must be a positive affine transformation of a weighted average of Dick's utility and Jane's utility. The set of permissible utility functions will be the set of all such weighted averages. It satisfies the convexity requirement on permissible utility functions which I advocated in "On Indeterminate Probabilities."[4] If Dick and Jane share the utility but have different probabilities, Pareto unanimity requires that every permissible probability be a weighted average of Dick's and Jane's. Again, the set of such permissible probabilities satisfies the convexity requirement.

Consider, however, the case where both the probabilities and the utilities of Dick and Jane differ in some respect. Seidenfeld, Kadane, and Schervish prove the remarkable result that the only two probability-utility pairs that preserve Pareto unanimity are those representing the views of Dick and of Jane. Hence, these are the only two permissible probability-utility pairs if we insist that Pareto unanimity is obeyed in consensus.

This result conflicts with the proposals I have made for such cases, according to which one should consider as permissible all weighted

on roulette lotteries over prizes and also over horse lotteries, a dominance or monotonicity axiom over horse lotteries and an axiom that renders each mixture of horse lotteries equal in value to a corresponding pure horse lottery, there is a unique probability p over the states and a state independent utility u unique up to a positive affine transformation such that the expected utilities of the horse lotteries according to (p, u) represent the weak ordering. In effect, the axioms ensure that the preference ordering satisfies the requirements of strict Bayes rationality when the states are probabilistically independent. If in addition to what is implied by the axioms, we assume that utilities are state-independent, strict Bayes rationality requires that expected utility be maximized according to the probability-utility pair given in the representation theorem.

4 *The Journal of Philosophy*, LXXI, 13 (July 18, 1974): 391–418. See also *The Enterprise of Knowledge* (Cambridge: MIT, 1980), and *Hard Choices* (New York: Cambridge, 1986). In the context of evaluating horse lotteries (on the assumption of state-independent utilities), a weak ordering of the horse lotteries is a permissible preference ranking of the horse lotteries if and only if there is a permissible probability-utility pair representing that ordering. A probability distribution over the states is permissible if and only if it is an element of some permissible probability-utility pair. The same for the permissibility of a utility function over the prizes. Given a subset of the horse lotteries as a set of feasible options, an element of the set is *E*-admissible if and only if it maximizes expected utility in the feasible set according to some permissible probability-utility pair. In my proposals, I require (a) that if p and u are permissible then so is the pair (p, u), (b) that the set of permissible probabilities is convex, and (c) the set of permissible utilities is convex. Seidenfeld, Kadane, and Schervish contend that these requirements are acceptable for consensual decision making only in certain restricted contexts.

Table 9.1

	s	$\sim s$
Option A	rl	rl
Option B	c_*	c

averages of Dick's probabilities and Jane's and all weighted averages of Dick's utilities and Jane's, and consider all methods of determining expected utilities to be permissible which are based on a probability and a utility permissible from these two sets.

There is no gainsaying the results demonstrated by Seidenfeld, Kadane, and Schervish. What may be questioned is whether unqualified Pareto unanimity is required in consensual decision making as these authors seem to think it is. The attractions of Pareto unanimity are undeniable. But there is a problem. To appreciate the nature of the problem, a numerical illustration (a minor variation of theirs) of their result may be useful.

Suppose Dick and Jane jointly face a choice between two "horse lotteries" in the technical sense of Anscombe and Aumann. Let c^*, c, and c_* be three prizes. Dick and Jane agree that c^* is better than c, which is better than c_*. Let rl be a roulette lottery that yields c with a chance of 19/30, c^* with a chance of 8/300, and c_* with a chance of 102/300. The two options faced by Dick and Jane are given in Table 9.1.

s and $\sim s$ are two "states" about whose truth Dick and Jane are uncertain. Dick assigns probability 0.1 to s (and 0.9 to $\sim s$) and Jane assigns probability 0.3 to s (and 0.7 to $\sim s$). Letting the utility of c^* be 1 for both agents and for c_* be 0 (as we may do without making any substantial assumption), Dick's utility for c is 0.1 and Jane's is 0.4. Computation reveals that the expected utilities assigned by Dick to option A and option B are both equal to 0.09. Jane's expected utilities for these two options are both 0.28.

Thus, in spite of the differences in their probability and utility judgments, Dick and Jane agree that A and B are to be equivalued.

Because of this agreement, Seidenfeld, Kadane, and Schervish conclude that the consensual point of view ought to preserve this equivaluation as well, just as Pareto unanimity requires. That is to say, any permissible comparison of A and B should rank them together.

We are assuming, however, that consensus is rational in the sense that every permissible ranking of the options should be representable by a pair of probability assignments to the states and utilities to the prizes such that the evaluation of the options is representable in terms of expected utilities.

Combining this requirement of Bayes rationality with Pareto unanimity implies that the expected-utility representations should rank *A* and *B* together. This implies that any permissible pair (p, u) of probability and utility functions should satisfy the following condition:

$$(19/30)u(c) + 8/30 = p(\sim s)u(c)$$

Let rl_1 be a roulette lottery yielding c^* with a chance 0.1 and c_* with a chance (0.9). Let the payoffs for options *C* and *D* be as given in Table 9.2.

Dick and Jane agree that *C* is not better than *D.* Pareto unanimity and Bayes rationality then require that the permissible probability-utility pairs should restrict probabilities to the interval between 0.1 and 0.3.

Let rl_2 yield c^* with chance 0.4 and c_* otherwise. Consider then the three options consisting of option *C,* option *E* which yields *c* for sure, and option *F* which yields rl_2 for sure. Dick and Jane agree that *C* is not better than *E,* which in turn is not better than option *F.* To preserve these shared agreements, the utility of *c* must be restricted to the interval from 0.1 to 0.4.

If we seek to preserve the shared agreements in preferences between *C, D, E,* and *F,* we now see that any probability-utility pair (p,u) which assigns probability to *s* in the range from 0.1 to 0.3 and assigns utility to *c* in the range between 0.1 and 0.4 will satisfy our requirements. If we seek to preserve the shared agreement concerning comparisons between *A* and *B* as well, however, the only probability-utility pairs from that set which satisfy our requirements are the pairs (0.1, 0.1) and (0.3, 0.4) representing Dick's and Jane's points of view, respectively.

An implication of this is that, given any pair of horse lotteries *H* and *H'* on the pair of states *s* and $\sim s$, such that Dick prefers *H* to *H'* and Jane prefers *H'* to *H,* there is no permissible evaluation of these options which ranks them together. There are no potential compromises.

Seidenfeld, Kadane, and Schervish generalize the result for this particu-

Table 9.2

	s	$\sim s$
Option *C*	rl_1	rl_1
Option *D*	c^*	c_*

lar example and show that, when two Bayes-rational agents differ in their probabilities and utilities over a set of horse lotteries, the set of potential resolutions that preserve shared agreements as Pareto unanimity does contains no compromises.

Seidenfeld, Kadane, and Schervish exhibit no doubt that unrestricted Pareto unanimity is a *conditio sine qua non* of consensus as shared agreement. They are prepared to abandon the inclusion of compromises among the potential resolutions. This means that, even when choosing between roulette lotteries (i.e., opportunities where chances are given), Dick and Jane must not countenance any compromise potential resolutions as long as they differ in both their probability and utility judgments and no matter how slight these differences might be. Consider, for example, a choice between receiving prize c for sure and receiving a roulette lottery that yields c^* with a chance of 0.25 and c_* with a chance 0.75. Dick prefers the roulette lottery to c but Jane prefers c to the roulette lottery. According to the position taken by Seidenfeld, Kadane, and Schervish, if Dick and Jane also have the option of flipping a fair coin to decide the issue, this option ought not to be counted as *E*-admissible. According to Dick's view, it is inferior to the roulette lottery. According to Jane's, it is inferior to the prize c.

Prohibiting such mixed options as inadmissible is surely an extreme view to take. I am not contending that adopting the mixed option is rationally obligatory. But it is not rationally prohibited either. Such prohibition is, however, an implication of the Seidenfeld, Kadane, and Schervish position if *E*-admissibility is necessary for admissibility.

To be sure, the Seidenfeld-Kadane-Schervish approach removes the prohibition if there is complete unanimity on either the probability or the utility function. It is a remarkable and, perhaps, objectionable aspect of their approach that it allows and, indeed, mandates the permissibility of compromises if there is complete agreement on either probability or utility, but prohibits compromises otherwise.

These reservations would, perhaps, be worth little were unrestricted Pareto unanimity a compelling requirement on consensus as shared agreement. And at first glance, it does seem obviously mandatory. What is a consensus as shared agreement if it does not preserve shared agreements?

A second glance, however, reveals that the case for Pareto unanimity is not only not compelling but is clearly untenable – at least for anyone who thinks that, in the evaluation of horse lotteries, "Bayesian" rankings representable by probability-utility pairs are the only potential resolutions in consensus. Since there is no disagreement between Seidenfeld, Kadane, and Schervish and myself on this score, we may leave this Bayesian assumption unquestioned.

Dick and Jane differ in their credal probabilities and utilities. Seidenfeld, Kadane, and Schervish do not tell us whether their differences are accompanied by a difference in their corpora (i.e., standards for serious possibility).[5] I myself do not think that the status of Bayes rationality and Pareto unanimity ought to depend on whether the participants disagree with respect to the corpus or not. In particular, either Pareto unanimity ought to apply in both cases or fail to apply in both cases. Whether Seidenfeld, Kadane, and Schervish agree is unclear. The fact that they overlook the case where participants in a joint decision differ with respect to corpus as well as with respect to probability and utility judgment suggests, however, that they may agree that the general conditions on consensus ought to cover both cases.

Consider then the case of Ron and Nancy, who resemble Dick and Jane in their credal probabilities and utilities as envisaged in our numerical example. Unlike Dick and Jane, it is clear that they also disagree with respect to the corpora to which they are committed as standards for serious possibility. Ron's corpus K_1 is the set of logical consequences of e and Nancy's corpus K_2 is the set of logical consequences of $\sim e$. Let us suppose, moreover, that Ron and Nancy agree in their confirmational

5 For discussion of serious possibility, see *Enterprise of Knowledge,* chs. 1-3, and *Decisions and Revisions* (New York: Cambridge, 1984, chs. 8, 9, 11, 12, and 14).

Characterizing *X*'s corpus of knowledge as *X*'s standard for serious possibility, as I have been accustomed to do, seems to be subject to a certain misunderstanding. Let us say that *X* fully believes that *h* when there is no serious possibility according to *X* that *h* is false. Then, according to what I have said about knowledge, everything *X* knows, *X* fully believes. What about the converse? From *X*'s point of view, everything *X* fully believes is true so that there is no difference between what *X* fully believes and what *X* knows. From *Y*'s point of view, however, there can be a substantial difference between what *X* knows and what *X* fully believes. Many critics have refused to appreciate my emphasis on this point.

Although I have characterized knowledge in terms of its function in inquiry, I have never offered this characterization as a necessary and sufficient condition for knowledge - except from the point of view of the knower. Is there a specification of necessary and sufficient conditions for knowledge which is the same for all agents? Yes! *X* knows that *h* if and only if *X* fully believes that *h* and it is true that *h*. A body of knowledge is, therefore, an error-free standard for serious possibility. From *X*'s point of view, *X*'s standard for serious possibility is error-free. The distinction between knowledge and full belief collapses as before.

What is ruled out is any appeal to a justificatory or explanatory condition on knowledge. Questions of justification arise when changes of point of view are being contemplated. With Charles Peirce, I say that, when there is no "living doubt," inquiry (and, hence, justification) is not demanded.

commitments. That is to say, Ron agrees with Nancy that, relative to K_2, 0.3 is the appropriate credal probability assignment to *s*. And Nancy agrees with Ron that, relative to K_1, 0.1 is the appropriate credal probability assignment. Finally, Ron and Nancy agree that, relative to the corpus *K* according to which both *e* and $\sim e$ are serious possibilities, the credal probability of *e* should be *x* and the credal probability of $\sim e$ should be $1-x$.

Under any familiar understanding of consensus as shared agreement, the parties to consensus should move to a corpus which preserves just those judgments of serious possibility and impossibility which they share in common, and which counts every proposition a serious possibility when parties to the consensus differ as to its status. Without agreement on the corpus, there is no consensus as shared agreement.

But if Ron and Nancy move to *K* and assign hypothesis *e* credal probability *x*, hypothesis *s* receives credal probability $0.1x+0.3(1-x)$. If we assume, as we consistently can, that neither Ron nor Nancy changes their utilities for roulette lotteries relative to *K* from what they were initially, then the probability-utility pair that represents consensus must rank options *A* and *B* in violation of unrestricted Pareto unanimity. Ron and Nancy both equivalue *A* and *B*. They both weakly prefer *F* over *E*, weakly prefer *E* over *C*, and weakly prefer *D* over *C*. As we have seen, the only probability-utility pairs that satisfy these consensual valuations are Ron's and Nancy's. On the other hand, as long as *x* is distinct from 0 and 1, the probability-utility pair representing consensus relative to the consensus corpus is distinct from either of these pairs. Options *C, D, E,* and *F* will continue to be ranked in the way required by Pareto unanimity relative to the consensus corpus, but *A* and *B* will not.

Observe that this objection to the Pareto-unanimity condition cannot be dismissed so readily as the implication that there are no compromises by saying that Pareto unanimity is definitory of the conception of consensus as shared agreement. Insisting that compromises of conflict be permissible does carry considerable plausibility. Still one may well doubt that, whatever the attractions of compromise may be, it is an ingredient in our understanding of consensus as shared agreement. But the requirement that consensus as shared agreement presupposes that potential resolutions of the conflicting views relative to the consensus corpus should be the permissible potential resolutions of the conflicting views of the parties to consensus relative to the corpora they actually adopt seems noncontroversial as a requirement on consensus as shared agreement. It requires that, in such consensus, the permissible probability-utility pairs should be defined over the space of possibilities determined by the consensus corpus. That should go without saying.[6]

Strictly speaking, I have begged the question against Seidenfeld, Kadane, and Schervish. But at this point, I must plead with the reader to consider whether a point of view is a consensus as shared agreement unless there is a single corpus relative to that consensus – the consensus corpus of shared agreements as to what is certain and unless the consensual evaluation of horse lotteries is relative to the consensus corpus.

To sustain my own affirmative answer to this question, however, it is desirable to appeal to more than presystematic conceptions of consensus as shared agreement, as I have done. After all, the Pareto-unanimity principle also seems presystematically compelling. In order to support my rejection of Pareto unanimity, it is desirable that a more modest version of the principle be suggested which can plausibly be said to accommodate what is presystematically attractive about Pareto unanimity.

Ron's corpus is K_1. Nancy's K_2. Ron can recognize that, were he to endorse Nancy's probability judgments relative to her corpus while keeping his utility function for roulette lotteries intact, his ranking of the elements of horse lotteries would differ from what it currently is. Some of the comparisons of horse lotteries would remain the same, however, relative to (0.3, 0.1) as they are relative to (0.1, 0.1). These comparisons of value we shall call Ron's robust preferences for the purpose of consensus with Nancy. Nancy like Ron has robust preferences for the purpose of consensus with Ron. With the notion of robust preferences for the purpose of consensus, we can then identify those robust preferences which Ron and Nancy share. We may speak of robust consensual strict preference, indifference and weak preference in a sense paralleling the corresponding notions of consensual preference. Finally, we obtain the idea of a robust consensual quasi ordering.

Observe that the robust consensual quasi ordering is induced by the

6 In the case envisaged, relative to the consensus corpus neither the probability-utility pair (0.1, 0.1) nor the pair (0.3, 0.4) is a potential resolution of the conflict. This does not imply that the differing points of view of Ron and Nancy are ignored. These two probability-utility pairs represent the viewpoints of Ron and Nancy relative to distinct corpora. We require only that their views relative to the consensus corpus be considered.

If these two probability-utility pairs were permissible, it would then turn out that, according to the consensus, the assignment of probability 1 to e and of probability 0 to e would both be permissible. This cannot be right relative to the consensus corpus. We have assumed that Dick and Jane agree that the credal probability for e is x relative to the consensus corpus. This might be 0 or it might be 1 (although neither is very plausible). But it cannot be both.

probability-utility pairs for Ron and Nancy relative to their current points of view whether they share the same corpus or not. We may distinguish that robust consensual quasi ordering from the robust consensual quasi ordering that they would endorse were they to move to the consensus corpus. We shall call the latter the *epistemically robust consensual quasi ordering.*

With these ideas in place, the Pareto-unanimity condition may be modified as follows:

Robust Pareto Unanimity: Every potential resolution is a consistent extension of the robust consensual quasi ordering over the set of horse lotteries.

Robust Pareto unanimity and Bayes rationality may now be taken to be necessary conditions on potential resolutions relative to the probability-utility pairs for the parties to the dispute, given the corpora they actually adopt. These two conditions are jointly sufficient.

If we assume that the probability judgments endorsed by Ron and Nancy relative to their current corpora are derivable from the consensus corpus by conditionalization and Bayes's theorem upon adding the information present in Ron's (Nancy's) current corpus but not in the consensus corpus, it is clear that all robust comparisons of horse lotteries relative to the consensus corpus are robust relative to the current corpora, though the converse does not hold except when Ron and Nancy share the same corpus.

Consequently, since Ron and Nancy do not share the same corpus, our current proposal does not imply that Ron's and Nancy's probability-utility pairs are permissible.

On the other hand, if Dick and Jane share the same corpus, our conditions do imply that both of their probability-utility pairs are permissible. Hence, these evaluations of horse lotteries are potential resolutions in this case. We get mutual respect for the views of the participants in consensual choice in the sense that these views count as permissible precisely when we want it - to wit, when their current corpora are one and the same and, hence, qualify as the consensus corpus (or where the probability-utility pairs relative to the consensus corpus are identical with those relative to the current corpora in all relevant respects so that disagreements in corpus are irrelevant).

What about the permissibility of compromises? Whether or not Ron and Nancy share a corpus in common, Bayes rationality and restricted Pareto unanimity imply that all weighted averages of the expected-utility functions defined by their probability-utility pairs represent potential resolutions. To see this, consider that *C* is robustly strictly preferred to *D* relative to Ron's and Nancy's views. Because of this, this preference is preserved by the pairs (0.1, 0.4) and (0.3, 0.1). Hence, they are also satisfied by all probability-utility pairs (p, u) where p is a weighted average of 0.1 and 0.3 and u is a weighted average of 0.1 and 0.4.

If a is strictly preferred to a' according to Ron and a' is strictly preferred to a according to Nancy, then either a and a' are equipreferred according to (0.1, 0.4) or according to (0.3, 0.1) or according to neither. If one of the first two alternatives obtained, the compromise condition is satisfied. If the third alternative is satisfied, then one of these pairs ranks a over a' or a' over a. Either way, there is a conflict between the ranking induced by this pair and either Ron's or Nancy's ranking. But now the conflict obtains between pairs that contain either a probability or a utility in common and, as we saw earlier on, the compromise condition is satisfied by the unrestricted Pareto unanimity and a fortiori the robust Pareto-unanimity condition in that case.

This does not mean that a compromise potential resolution will be permissible in consensus. For this to obtain, either Ron and Nancy must share the same corpus or, if they have different potential corpora, the set of potential resolutions of their conflicts relative to the consensus corpus must include compromise potential resolutions relative to the actual corpora for each paired comparison between horse lotteries. This need not, in general, obtain.

Thus, it has been shown that robust Pareto unanimity is compatible with the requirement that there be permissible compromises of conflict even though the unqualified Pareto-unanimity condition is not. It has also been shown that robust Pareto unanimity is compatible with the requirement that, in consensus as shared agreement, judgments should be made relative to the consensus corpus. Unqualified Pareto unanimity is not compatible with this requirement.

Seidenfeld, Kadane, and Schervish, in effect, point out that weakening Pareto unanimity to robust Pareto unanimity fails to make a distinction between seeking consensus as shared agreement when Dick and Jane are the participants, when Tom and Mary are the participants and where Tom's evaluation of the given options is representable by (0.1, 0.4) and Mary's by (0.3, 0.1), and when Dick, Jane, Tom, and Mary all join it. In these three cases, the pairs of acts about whose ranking all parties to consensus agree are different. But the differences occur only with respect to those agreements which lack robustness. With respect to robust agreements, all three cases are the same and, hence, the set of potential resolutions on my view are the same. Seidenfeld, Kadane, and Schervish seem to think that this is a defect in my view. It fails to make fine-grained enough distinctions. They see vice in neglecting nonrobust consensual preferences in consensus. Where they see vice, I see virtue.

To see the virtue, suppose that Ron, Nancy, George, and Barbara agree that 0.1 is appropriate to use as the probability of *s* relative to *e* and 0.3 is appropriate to use relative to ~*e*. Ron and George agree that *e* is true and Nancy and Barbara that ~*e* is true. Now, just as Ron and Nancy should reach consensus as shared agreement by moving to suspense between *e* and ~*e*, so should George and Barbara and so should all four do when they are seeking to make a joint consensual decision. If they all agree that *x* is the probability of *e* in the state of suspense, the consensus probability-utility pairs will be the same in all three cases. If they recognize all values of *x* in some range as permissible, the same will be true. In short, there will be no difference between the three cases of consensus when the parties do not share a common corpus to begin with. If the status of Pareto unanimity is the same whether participants to a consensus share the same corpus or not, as I am maintaining, there should be no difference between the cases of consensus when the parties do share a common corpus at the outset.

I concede this much to the unqualified devotion to Pareto unanimity advocated by Seidenfeld, Kadane, and Schervish. A Catholic and a Communist might face a choice between two policies and agree that one is better than the other even though they differ in probability and utility. They might agree to act in concert without seeking a consensus as shared agreement. They might rest content with a political consensus. But such a political consensus is not consensus as shared agreement. Indeed, the Catholic and Communist both view the joint action as the action of two agents each acting separately to promote his or her personal agenda as in a game. Both prefer a national health-insurance plan with a prohibition of aid for abortions to no health plan at all. Each will vote for the plan. But their both voting is not a group action grounded on consensus as shared agreement. The alleged group action is the implementation of the national-health plan which by hypothesis is the outcome of their both voting for it. In our example, it turns out that a form of Pareto unanimity will be observed. The Catholic and the Communist both prefer the health plan being implemented than not. And their "votes" will result in its implementation. But as is well-known, there are game-like situations (e.g., prisoners' dilemmas) where the "votes" or choices of the participants will yield Pareto-inferior outcomes even though the players are fully rational. In any case, when group action is the outcome of the independent decisions of members of the group, there is no failure of group rationality. There is no group agent to fail.

Seidenfeld, Kadane, and Schervish start with the premise that the kind of group decision making they are considering is not of the game-like sort just mentioned but belongs to the category of social-choice problems where the decisions made should meet canons of rationality applicable to individual choice. Yet, the result they come up with according to which the only resolutions of the conflict between Dick and Jane are the views of Dick and Jane is, for reasons I have already indicated, unacceptable as an account of consensual decision making. It remains an open question whether there are any contexts of group behavior where their requirements may plausibly be imposed.

Seidenfeld, Kadane, and Schervish have produced an impossibility theorem which is at once a surprising and novel contribution to the Arrovian tradition and a fertile source for philosophical reflection. We should be grateful to them for having opened up new avenues for exploration in theories of individual and group decision making. Nonetheless, I think their unrestricted endorsement of Pareto unanimity and abandonment of the requirement that in consensus compromises should be countenanced as potential resolutions is misguided. To my way of thinking, the important lesson to be drawn from their theorem is that adherence to Pareto unanimity is not all it is cracked up to be.

10
The paradoxes of Allais and Ellsberg

PRESCRIPTIVE AND DESCRIPTIVE ADEQUACY

In *The Enterprise of Knowledge* (Levi, 1980a), I proposed a general theory of rational choice which I intended as a characterization of a prescriptive theory of ideal rationality. A cardinal tenet of this theory is that assessments of expected value or expected utility in the Bayesian sense may not be representable by a numerical indicator or indeed induce an ordering of feasible options in a context of deliberation. My reasons for taking this position are related to my commitment to the inquiry-oriented approach to human knowledge and valuation favored by the American pragmatists, Charles Peirce and John Dewey. A feature of any acceptable view of inquiry ought to be that during an inquiry points under dispute ought to be kept in suspense pending resolution through inquiry. I contend that this sensible attitude ought to be applied to judgments of probability and value or utility. This consideration ought to lead to a form of indeterminacy in probability judgment, utility judgment and assessments of expected utility.

In Levi, 1980a, I did not examine the extent to which the proposals I made accommodate or fail to accommodate the results of empirical investigations exploring the rationality of probability judgment, utility judgment and decision making which have mushroomed in recent decades.

In this paper, I shall partially remedy this defect by examining the way the proposals under consideration address the so-called paradoxes of Allais and Ellsberg, the "preference reversal" phenomena uncovered by Grether and Plott, and responses to the Newcomb problem. I shall show that the theory of rational choice found in Levi, 1980a, can give accounts of responses of experimental subjects in a systematic manner.

The relation between prescriptions regulating rational belief, rational preference and valuation, and rational decision making, on the one hand, and what empirical studies reveal about the judgments and conduct of

From *Economics and Philosophy* 2 (1986), pp. 23–53.

deliberating agents, on the other, is a troubling one. It is always open to someone defending a conception of ideal rationality to insist that empirical evidence indicating deviation from the prescriptions favored by his or her recommendations proves only the urgency of educating stiff-necked humanity to think and deliberate better. Determined empiricists, by way of contrast, may contend that prescriptions which are systematically and pervasively violated by human agents are useless even in their capacity as norms and ought to be abandoned or tailored to the empirical data.

I think the truth lies somewhere between these extremes. It is quixotic to seek to construct a system of prescriptions for coherent and rational decision making which human agents can always and automatically obey. No matter how attractive a system of norms may be and no matter how easy it is relative to rival systems to make the computations requisite for its application, it is to be expected that situations will arise where the calculations needed for applying the system will be sufficiently complex to outstrip the capacity of human agents unaided by assorted technologies such as books, pencil and paper, computers, and mathematical approximations. And even with such crutches, problems may arise where the resources of the computational technology will be stretched beyond their limits. This general consideration ought to suggest that the extreme empiricist attitude is excessive in its demands. We ought to be prepared to settle for less than a system of norms which are applicable by human beings on all occasions and which they may be expected to apply with a high degree of regularity.

On the other hand, it would be a mistake for students of norms of ideal rationality to ignore data on actual decision making altogether. It is one thing to point out that human beings have limited computational capacity, limited memory, tend to ignore subtle differences, and are often distracted by emotional upset of one sort or another, and that such limitations can and do lead to deviations from ideal rationality. It is quite another to suppose that all deviations from some system of norms can be attributed to failures of these kinds. It will, indeed, often be difficult to decide whether failures of computational capacity or emotional upset adequately account for failures of rationality. Still on occasions where the agent insists on going his or her own way in spite of apparently successful efforts having been made to explain to the agent that he or she is deviating from the norms of rationality under consideration, we should conclude that the agent, at any rate, refuses to subscribe to the norms. If such refusal is sufficiently widespread, the adequacy of the norms as universal ideals should be open to question. Rather than looking for methods to encourage delinquent individuals to change their ways, it may be worth asking whether the ideals of rationality ought to be modified to render the wayward agent respectable. Even here, theoretical arguments in favor of retaining the ideals should sometimes win the day; but the pressure to revise the ideals ought to be greater when deviations from them cannot be attributed to lack of memory, computational capacity, inattention, emotional instability, and the like than when such explanations seem more readily available.

I acknowledge at the outset that the considerations I have mentioned, which excuse those who study prescriptive ideals of rationality from accommodating data concerning deviation from their norms, are vague. I cannot myself make them more precise. One may hope that as investigations of these matters in psychology, economics, and the other social sciences proceed, we may improve upon these vaguely specified excuses in a fruitful way. For the present, I shall remain content to assume that this vaguely drawn distinction can be made between deviations from norms of ideal rationality which call for behavior modification (rather than revision of the ideals) and deviations which call for modification of the norms themselves.

In addition to offering my own analysis of the various "paradoxes" of rational choice, I shall consider other approaches to the puzzling phenomena noticed by Allais and Ellsberg. It will be argued that none of these approaches can accommodate all these cases in the way the scheme I shall present can and, moreover, that these proposals have unfortunate consequences which even their own advocates ought to find disquieting.

None of this proves that the proposals I favor are superior to their competitors. Efforts at proof are, I fear, futile. But if these proposals do better than rival schemes in accommodating the puzzling cases, that should surely be considered an argument in their favor.

Along the way, I shall point out one variant of the puzzles generated by Allais which the theory under examination cannot digest. My own suspicion is that this particular type of case is an example of a deviation from rationality which may be attributed to inattention and failure to calculate and, hence, ought to be counted as human failure rather than a failure of the ideals of rationality being proposed.

This claim will no doubt be disputed. Given the fact that the proposals offered here accommodate a much wider variety of the phenomena canvassed than the available alternatives (at least those with which I am familiar), the balance of reasons remains in their favor – pending the construction of a more comprehensive approach or a strong argument showing that my explanation of the indigestibility of the case in question is untenable on empirical or theoretical grounds.

Several features of the proposals advanced here may provoke strong dissent. I shall take note in passing of deviations from conditions on consistency of choice and from conceptions of the relation between choice and preference which lie at the core of much thinking in economics and decision theory. Where I suspect others will see vice, I often see virtue or, if not virtue, nothing vicious. Given the systematic generality of the theory, its range of applicability, and its attractiveness from the vantage point of the inquiry-oriented approach to knowledge and value I borrow from Peirce and Dewey, my dissent from orthodoxy on these matters represents something more than wilful perversity. My hope is that others whose antecedent commitments differ from mine will recognize the generality of the theory as sufficient grounds for giving it a serious hearing in spite of its alleged warts and blemishes.

RISK AND UNCERTAINTY

In chapter 2 of their justly celebrated *Games and Decisions,* R. D. Luce and H. Raiffa (1958) classify decision problems into three categories: decision making under certainty, under risk, and under uncertainty (p. 15). Decision making under certainty obtains if "each action is known to lead invariably to a specific outcome." Risk obtains "if each action leads to one of a set of possible specific outcomes, each outcome occurring with a known probability," and uncertainty obtains "if either action or both has as its consequence a set of possible specific outcomes, but where the probabilities of these outcomes are completely unknown or are not even meaningful."

When Luce and Raiffa speak of known or unknown probabilities, they intend objective or statistical probabilities. According to their taxonomy, if an agent does not know the objective probabilities of outcomes of a given action but, nonetheless, assigns definite numerical probabilities to the possible outcomes of an action given that it is implemented, which represent his or her degrees of belief or confidence, the decision problem remains an instance of decision making under uncertainty. Thus, they discuss the proposal of L. J. Savage to rely on numerical personal probabilities in these circumstances under the rubric of decision making under uncertainty rather than under risk (p. 300).

Luce and Raiffa themselves acknowledge that Savage's personalism "reduces the decision problem from one of uncertainty to one of risk" (p. 300), confessing thereby to a certain confusion in their taxonomy. Savage favored maximizing expected utility where the expectations are calculated using the decision maker's personal or credal probabilities. In decision making under risk, as construed by Luce and Raiffa, one should also maximize expected utility. But here the expectation-determining probabilities are alleged to be objective, statistical probabilities or chances or, more accurately, the statistical probabilities the decision maker assumes, takes for granted, or is certain to obtain. Luce and Raiffa cannot quite make up their mind as to whether to stick to this characterization or to regard decision making under risk as present whenever it is appropriate to maximize expected utility, regardless of the status of the expectation-determining probabilities.

When an agent is certain that a coin is fair and is about to be tossed and knows nothing else relevant to the outcome of the toss, his personal degree of belief that the coin will land heads on that occasion ought to be 0.5. That is the noncontroversial core of principles of direct inference (Levi, 1980a, ch. 12). If offered a gamble where he wins S utiles on heads and wins nothing otherwise, the agent ought to be willing to pay up to 0.5S utiles for the privilege (assuming neither taste nor aversion for gambling). Thus, the probabilities used to compute expected value are equal to the statistical probabilities the agent is certain to obtain; but we can with justice regard these expectations determining probabilities as personal or credal probabilities, as per Savage. The only difference is that these personal probabilities are grounded in knowledge or certain conviction in the truth of assertions about statistical probabilities.

Savage, of course, followed de Finetti in rejecting statistical probabilities (when not reduced to personal probabilities) as metaphysical moonshine, although he was more polite than de Finetti in saying so (Savage, 1954, p. 54). According to Charles Peirce, however, it is metaphysically preposterous to suppose that worlds are as plentiful as blackberries. He is best understood as asserting that we are entitled to judgments of numerically definite credal probability usable in computing expected value only when these can be justified by direct inference from knowledge of statistical probabilities (Levi, 1980b). So are the modern founders of statistical theory, R. A. Fisher, J. Neyman, and E. S. Pearson. (See Fisher, 1959, pp. 31-35, and Pearson, 1962, p. 277.) In spite of their opposition to Bayesianism these authors agreed with Savage that if one is in a situation where there is a warrant for making numerically definite credal probability judgments, one should regard these probability judgments as appropriate to use in computing expected value or utility and, in assessing policy, should maximize expected utility relative to such credal probabilities.

To be sure, Savage differed from Fisher and Neyman with respect to the scope of applicability of the principle that expected utility is to be maximized. But according to the reconstruction I have just briefly outlined, that scope coincided, for all of them, with the range of legitimate judgments of numerically definite credal probability. Their disagreements boiled down to differences in their views as to the conditions under which numerically definite credal probability judgments are legitimate. Peirce, Fisher, Neyman and Pearson thought these conditions are far more circumscribed than Savage did.

Luce and Raiffa restrict the category of decision making under risk to cases where numerically definite credal probabilities can be derived from knowledge of simple statistical hypotheses specifying definite objective probability distributions.

But since the very intelligibility of objective probability is a matter of controversy and, even among those who agree that it is intelligible, the conditions under which it is legitimate to make numerically definite credal probability judgments remain problematic, it does not seem sensible to characterize decision making under risk in a manner which makes the importance of the category stand or fall with the cogency of a controversial outlook in probabilistic epistemology.

For this reason, I have tended to think of decision making under risk as arising in any context where an agent legitimately assigns numerically definite credal probabilities to hypotheses about the outcomes of feasible options. In other words, decision making under risk is any context in which maximizing expected utility is the appropriate criterion for identifying admissible options.

On this view, decision making under uncertainty arises when it is illegitimate to make numerically definite credal probability judgments. On Savage's theory, it is always legitimate to make such judgments. There is no such thing as decision making under uncertainty.

More important, this way of presenting things emphasizes the correct insight of the Luce and Raiffa approach that when we are entitled to make numerically definite credal probability judgments legitimately, appraisals of feasible options with respect to expected utility or value ought to take precedence over alternative valuations of the feasible options. I do not pretend to have a proof of this claim. Any alleged proof will seem to the determined skeptic like question begging.

Fortunately there is a broad consensus among both Bayesians and many non-Bayesians that when it is legitimate for an agent to make numerically definite credal probability judgments about the outcomes of the options available to him, he should evaluate those options with respect to expected utility. The existence of this consensus indicates that defending the priority of considerations of expected utility is not a pivotal issue in much of the criticism of Bayesian decision theory.

The chief bone of contention between Bayesians and many anti-Bayesians concerning both statistical theory and decision making has been the conditions under which it is legitimate for agents to assign numerically definite credal probabilities appropriate for computing expected utilities to hypotheses.

No doubt there is another anti-Bayesian tradition, primarily spawned by economists rather than statisticians, who have been skeptical of the explanatory power of the expected utility hypothesis even when probability judgments may be made with numerical definiteness. Thus, Alfred Marshall, who assumed that the marginal utility of income diminished as income increased, regarded willingness to accept even money bets as the product of "pleasure derived from the excitement of gambling - a pleasure likely to engender a restless, feverish character unsuited for steady work as well as for the higher and more solid pleasures of life" (Marshall, 1920, n. 9, p. 694). Marshall's view has often been understood as an acknowledgment that agents fail to maximize expected utility even when probabilities are numerically definite and, indeed, was so construed by M. Friedman and L. J. Savage (1948, pp. 280–81). I have not myself understood why an agent who accepts even money bets because of a taste for gambling even though the marginal utility of income decreases fails to be an expected-utility maximizer,[1] but, for better or worse, interest in phenomena of risk aversion and risk attraction, among other things, has persuaded some authors that even when probabilities are numerically definite, expected utility may fail to be maximized.

In recent years, the injunction that expected utility be maximized has also been criticized by advocates of some form of causal decision theory

1 Having either a taste or an aversion for gambling is consistent with being an expected-utility maximizer. However, there is then no well defined utility of money function. An extra dollar received for sure and an extra dollar received as the outcome of gambling increase utility by different amounts. Hence, strictly speaking, the principle of diminishing marginal utility of money is meaningless. But if we restrict its scope to the utility of money received other than by gambling, perhaps its intelligibility can be saved. When so restricted, the principle of diminishing marginal utility of money and the expected utility principle may both be satisfied by someone who accepts even money odds on 50–50 gambles. Some theorists may prefer to retain the idea of a utility of money function and sacrifice the expected utility hypothesis.

which favors invoking a new version of the principle that expected causal utility be maximized (Gibbard and Harper, 1978).

In my judgment, those cases where people appear to be deviating from the injunction to maximize expected utility, and where it takes a heavy dose of Bayesian dogmatism to insist that the agents are behaving unreasonably, can by and large be understood as situations where either credal probability judgments which are numerically definite cannot legitimately be made, or utility assignments unique up to a positive affine transformation are inappropriate. It is indeterminacy, and not risk aversion or causal structure, which is the salient feature of those contexts where the applicability of Bayesian decision theory breaks down.

When situations arise where the principle of maximizing expected utility cannot render a verdict as to what is optimal or admissible because numerically definite credal probability judgments cannot be acceptably made, we may look to some principle of decision making under uncertainty as a secondary criterion to use. This much I have adopted from the outlook of Luce and Raiffa as just amended.

However, the vision Luce and Raiffa offer us suggests that the difference between decision making under risk and under uncertainty is an all-or-nothing affair. Either we are entitled to assign numerically definite credal and expectation-determining probabilities and to maximize expected utility relative to them or we are not entitled to make any credal probability judgments at all.

The tendency to see matters in this way spills over to discussions of risk management and public policy where disputes arise as to whether one should take into account expected costs and benefits or whether one should fixate on worst possible cases. (For a discussion of this polarization in connection with assessing the safety of nuclear power plants, see Levi, 1980a, appendix.)

Abraham Wald is one of the many who insisted that if a probability distribution "existed," the option bearing maximum expected utility (or, since Wald spoke in terms of "risk," the option bearing minimum expected risk) is an "optimum" (Wald, 1950, p. 16). On the other hand, if a probability distribution does not "exist" or "is unknown to the experimenter," Wald regarded "minimax solutions" recommending minimizing the maximum possible risk as "reasonable" (Wald, 1950, p. 18).

Wald sought to determine the conditions which a decision problem should satisfy under which minimax solutions are also "Bayes solutions" – i.e., maximize utility or minimize expected risk relative to some definable probability distribution regardless of whether such a distribution "exists" (characterizes some stochastic process). (See especially Wald, 1950, ch.

3.) This seems to suggest that, for Wald, under the "general conditions" he describes for decision making under uncertainty, all probability distributions over the given space of consequences or states are permissible to use in computing expected utility. That is to say, we should recognize as admissible with respect to expected utility any option which comes out best in expected utility relative to some such distribution. Rather than say that the agent makes no probability judgment, I prefer to say that the agent makes no numerically definite credal probability judgment. But he does make credal probability judgments. In decision making under uncertainty, he refuses to rule out any consistent distribution from being permissible to use in computing expectations, whereas in numerically definite probability judgments, he rules out all but one distribution.

When matters are put this way, the dichotomy between decision making under risk and under uncertainty gives way to a more finely grained set of distinctions. There can be intermediate credal states between numerically definite credal probability judgments and the maximally indeterminate ones that characterize decision making under uncertainty. A credal state for propositions belonging to some suitably specified space of possibilities may be represented by a single credal probability distribution (the uniquely permissible one), the set of all distributions over that space (the maximally indeterminate credal state) or by any convex subset of that maximally indeterminate set (Levi, 1980a, chs. 4.1, 5, and 9).

IMPRECISION AND INDETERMINACY: THE ELLSBERG PROBLEM

There is a widespread tendency to confuse indeterminate credal probability judgments of the sort I am describing with forms of self-ignorance – i.e., the ignorance of one's own numerically definite credal probability judgments. This tendency is especially prevalent among Bayesians, who tend to the view that at least ideally rational agents ought always to have numerically definite credal states. In the face of the apparent fact that perfectly sensible agents refuse to make numerically definite judgments, such strict Bayesians tend to see the lack of numerical definiteness as the product of a measurement problem. They are prepared to acknowledge such indefiniteness in much the same spirit that they will acknowledge that length may be measured by a certain measuring stick to the nearest thirty-sixth of an inch. To report that the width of a table as measured by such a stick is three feet ought more carefully to be a report that the width is three feet plus or minus one seventy-second of an inch. In this way, we express our ignorance as to the true unknown precise value of the width if we think there be such. So too, indefiniteness in credal probability is supposed to represent our ignorance as to our unknown strictly Bayesian credal state.

According to the view I am proposing, credal states may be genuinely indeterminate. An agent who is fully in touch with his views may insist that he regards all distributions in a given set as permissible to use in computing expectations rather than that he is ignorant as to which of the distributions in a given set is the uniquely permissible one to use in computing expectations. The failure to single out a unique distribution cannot be remedied by more precise techniques of measurement.

The difference to which I am pointing may be approached from a somewhat different angle - namely, the ramifications of indeterminacy for decision making.

Consider the following problem posed by Daniel Ellsberg (Ellsberg, 1961). We are given an urn containing 30 red balls and 60 balls which are white or blue in unknown proportions. The decision maker is offered the following hypothetical choice with the reward contingent on the color of a ball drawn from the urn.

	Red	*White*	*Blue*
A	100	0	0
B	0	100	0

Here the decision maker knows that the objective probability of obtaining a red ball is $\frac{1}{3}$ and that the objective probability of obtaining a white is anywhere from 0 to $\frac{2}{3}$ in increments of sixtieths. If the agent's credal state concerning the hypotheses as to the precise contents of the urn (there are 61 hypotheses as to the precise proportions of white balls in the urn) is maximally indeterminate, there are some permissible distributions according to which the expected value of choosing A is greater than that of choosing B and some according to which the ranking is the other way around. (There is, of course, one according to which they are equal in value.) Both options are admissible with respect to expected utility (*E-admissible*). We have a decision problem under uncertainty and may invoke a secondary criterion.

The criterion I favor using is a maximin criterion or rather a lexicographical maximin. If one speaks of losses rather than benefits or utilities, the criterion is a minimax or lexicographical minimax criterion.

Ellsberg had alleged that using maximin would be of no help in this particular example. The "worst possible case" is 0 for both options A and B and the second worst case is 100. Maximin cannot decide - or so it seems.

But there are many ways to compute worst possible cases. The method favored by Abraham Wald considered a case to be distinguished according to which hypothesis concerning the precise contents of the urn is true. In this sense, there are 61 possible cases in the Ellsberg problem corresponding to each of the 61 hypotheses specifying how many of the sixty non-red balls are white and how many are blue. One can compute, for each such case, the expected value of an option conditional on the truth of that case and then assess security levels. Indeed, when the payoffs are represented as losses, these expectations are precisely what Wald called "risks" (Wald, 1950, p. 12): No matter which of these 61 hypotheses is true, the expected value of A is 100/3. This value and not 0 is now regarded as the security level for A because it is the conditional expected utility payoff for each of the 61 cases and hence is the worst possible payoff among these 61 cases. Turning to option B, the conditional expected utility payoff when 60 balls are white and none blue is 200/3. The payoff declines as the number of white balls is reduced and becomes 0 when all 60 balls are blue. The security level for B is therefore 0. Whereas the way we calculated security levels originally assigned equal 0 security levels to both A and B, according to the Wald approach the security level associated with option A is 100/3 and the security level associated with option B is 0. Option A becomes uniquely admissible.

I do not think there is any way of deciding which method of determining security levels is preferable or rational. That question ought to be left up to the decision maker and is to be regarded as a value commitment on the agent's part. Still it is rationally acceptable for a decision maker to follow Wald's method and to choose A because it is the maximin solution. Although Ellsberg was right in observing that the first way to use maximin cannot decide between A and B, Wald's way of doing so does. Hence, if we find a decision maker who chooses A over B, and if we assume that his credal probability judgments are indeterminate in the manner indicated, we may infer, according to the model I am proposing, that he is fixing security levels after the fashion of Wald. Someone else who regards both A and B as admissible should be judged to be assessing security in another way - perhaps as Ellsberg suggests. Neither agent is irrational. They differ in their concern for security.

But now consider the following decision problem:

	Red	*White*	*Blue*
C	100	0	100
D	0	100	100

As before, both options are *E*-admissible. If we follow Wald's method, the security level for C is 100/3 as before; but the security level for D is now 200/3 so that D should be chosen. Again one does not have to fix security levels in this fashion. One could fix them so that both options remain admissible, but it is rationally acceptable to regard D as uniquely admissible. And, indeed, the agent who regards A as uniquely admissible in the first problem should regard D as uniquely admissible in the second problem.

Now the strict Bayesian thinks of the lack of definiteness in the representation of the agent's credal state as due to imprecision in measurement, and insists that the ideally rational agent is committed or ought to be committed to a uniquely permissible distribution even if the agent cannot tell precisely what that distribution is. Such a strict Bayesian will insist that the choice of A over B and D over C is rationally untenable. It is acceptable to regard A as uniquely admissible in the first problem and C in the second or B uniquely admissible in the first and D in the second. And one might regard both as admissible in each problem. But what the Wald theory recommends here is utterly forbidden. It violates the "sure-thing principle."

The sure-thing principle states that if an agent were to consider two distinct decision problems which had identical payoff structures, except that relative to one column in the first all payoffs were given the value *a*, whereas in the corresponding column in the second problem all payoffs were given the value *b*, then, if A is strictly preferred to B in the first, it is in the second as well (Savage, 1954, pp. 20–22).

The strict Bayesian thinks that the recommendations I have allowed to be reasonable violate the sure-thing principle; for the strict Bayesian supposes that the decision maker is committed to behaving as if he had a numerically definite credal state whether he can tell what it is or not, and hence that the agent is committed to ranking the options in a manner which yields a weak ordering.

But on the view I favor, considerations of expected utility fail to yield an ordering of the options. The options are noncomparable with respect to expected utility. Since A is not preferred to B in the first problem (nor, for that matter, is it indifferent to B), there is no violation of the sure-thing principle.

What is true is that the use of the Wald criterion leads to choosing an option which is best according to one subclass of permissible distributions in problem 1 (those for which the probability of blue is no less than $\frac{1}{3}$) and another subclass in problem 2 (those for which it is less than $\frac{1}{3}$).

The strict Bayesian thinks this is incoherent or wishful thinking. One cannot modify one's credal probability judgment simply because the payoff matrix has changed.

But on the view I favor, there has been no modification of the credal state. It is indeterminate in both problems and in the same way. The same set of distributions is permissible to use in computing expectations. And in both problems considerations of expected value fail to render a verdict. The use of the Wald version of the maximin criterion is not designed to single out a more determinate credal state relative to which one will maximize expected utility. Rather considerations of expected utility having failed to render a verdict, considerations of security are brought into play. To do this is not to endorse a pessimistic or paranoid credal state in lieu of the one with which one initially began. It is not to modify the credal state from its pristine indeterminacy at all.[2]

Care must be taken in understanding the force of this account of the Ellsberg problem. It is supposed here that the agent is offered the first choice in a context where he takes for granted that he is not being offered the second and is offered the second in a context where he takes for granted that he is not being offered the first. Of course, no real-life decision maker with a minimal amount of consciousness and memory can be placed in both contexts. The responses we have been considering are hypothetical reactions to hypothetical scenarios.

If the decision maker were offered both pairs of options simultaneously or in some temporal order and recognized this to be the case, he should address both problems together. He may then be taken to have four options: choose A and C, A and D, B and C and B and D. Each of the four options turns out to be *E*-admissible. But only the joint choices of A and D and B and C are admissible when considerations of security are brought into play. Yet, when the two problems are considered separately, one cannot regard B as uniquely admissible in the first problem and C as uniquely admissible in the second.

Suppose one is offered a 50–50 mixture of A and D and a 50–50 mixture of B and C. The two "compound" lotteries reduce to the same simple lottery and, hence, should be regarded as indifferent to one another. This allegedly violates an "independence" postulate because A is preferred to B and D to C yet a mixture of A and D is indifferent to a mixture of B and C (Raiffa, 1961, p. 694). Once more the charge is false. There is no such violation; for A is not preferred to B and D is not preferred to C. There is no *preference* one way or the other in either case.

2 Theorem 3.9 of Wald, 1950, asserts that under certain conditions any minimax solution to a decision problem is a Bayes solution relative to a "least favorable" probability distribution. This rhetoric has tended to foster the view that advocacy of minimax or maximin is a species of pessimism or paranoia.

Thus, the account I have given using the Wald version of the maximin principle as a secondary criterion of admissibility when considerations of expected utility fail to render a verdict manage to yield the verdicts in the Ellsberg problem which Ellsberg reports to be prevalent among those who address it. At the same time, counter to the arguments of staunch defenders of the Bayesian faith, like Raiffa, endorsement of these recommendations does not lead to absurd recommendations when we confront a choice between a mixture of A and D and a mixture of B and C. There is no violation of the sure-thing principle. The so-called independence postulate remains intact. Every admissible option is a Bayes solution according to the credal state.

Recently Peter Gärdenfors and Nils-Eric Sahlin have suggested another approach to the Ellsberg problem (Gärdenfors and Sahlin, 1982, pp. 374–377; Levi, 1982b, pp. 401–408). They too begin with a set of probability distributions to represent the state of probability judgment. They compute for each option its lowest expectation value and recommend maximizing the lowest expectation value. Elsewhere in the statistical literature this has been called "gamma minimax." (See Berger, 1980, pp. 134–135, who credits Robbins, 1964, with the first explicit formulation of the idea.)

Gamma minimax seems to me to be objectionable for two reasons:

1. It fails to require every admissible option to be a Bayes solution – i.e., to be best in expected utility according to some permissible probability. This objection may not appear decisive to those who mean to challenge the primacy of *E*-admissibility as a test of admissibility. But as noted earlier, many critics of strict Bayesian doctrine do not appear to object to the primacy of *E*-admissibility. They ask rather where the numbers come from. Wald, who was no Bayesian, was quite anxious to identify the conditions under which minimax solutions are Bayes solutions. The objection is serious even if it is not decisive.
2. One of the advantages of insisting on the primacy of *E*-admissibility is that it provides insurance against recognizing as admissible an option which is dominated by another option in the admissible set. Gamma minimax lacks the virtue. Teddy Seidenfeld has constructed an argument which shows that gamma minimaxers can be inveigled into choosing options dominated by other feasible options.[3]

3 Imagine a shell game with two shells: the left and the right. According to option L, the agent wins $100 if there is a peanut under the left shell and wins $10 if the peanut is

The failure of gamma minimax to yield a Bayes solution does not emerge in the Ellsberg problem. It can be seen in the following example:

Suppose an urn contains black and white balls in an unknown proportion ranging from 40% black to 60% black. Three decision problems are then considered:

Case 1: The agent is offered a gamble where the agent wins 55 utiles if a black is drawn and loses 45 utiles if a white is drawn. He can either accept the gamble or refuse it and neither win nor lose.

Case 2: The agent is offered a gamble where the agent wins 55 utiles if a white is drawn and loses 46 utiles if a black is drawn. He can either accept the gamble or refuse it and neither win nor lose.

Case 3: The agent can accept the gamble from case 1, the gamble from case 2 or refuse both. He must choose one and only one of these options.

In case 1, both options are *E*-admissible, but refusing to gamble brings greater security so that refusal is uniquely admissible.

In case 2, both options are *E*-admissible, but refusal brings greater security and is uniquely admissible.

In case 3, refusal is not *E*-admissible, but the other two options are. Since of the two gambles, the security level in case 1 is higher, that option is uniquely admissible.

under the right shell. According to option R, the agent wins $100 if the peanut is under the right shell and $10 if the peanut is under the left shell. The largest price a gamma minimaxer would be prepared to pay for either gamble is $10. Consider now a lottery where there is a 50–50 chance of receiving gamble L or gamble R. The gamma minimax value in utiles for this gamble is equal to the expected utility of a 50–50 gamble for $100 or $10 and this will have a dollar gamble greater than $10. Let this value be $25. Consider now a choice between the following pair of lotteries: According to lottery 1, if a fair coin lands heads up, the agent has a choice between gamble L and $9 and if the coin lands tails up, the agent has a choice between R and $9. Because, according to gamma minimax, L is preferred to $9 and R to $9, lottery I is equivalent to the 50–50 chance of receiving L or R and this has a value of $25. According to lottery II, if the coin lands heads up, the agent has a choice between L and $11 and if the coin lands tails up, he has a choice between R and $11. If lottery II is taken, the agent will choose the money according to gamma minimax regardless of what happens. Hence, the gamble is worth $11. Hence, the gamma minimizer will choose lottery I over lottery II. Consider, however, that if the coin lands heads up, the dollar value of I is $10 while that of II is $11. The same obtains if the coin lands tails up. Option II dominates option I while gamma minimaxers such as Gärdenfors and Sahlin favor option I. According to the proposal I made, by way of contrast, both options are as admissible in a choice between $10 and gamble L. But that does not mean that they are of equal value or utility. The same holds for a choice between R and $10. Indeed, in a choice between L and $11 for sure, both options are *E*-admissible. The same holds for a choice between R and $11. Hence, option II does not dominate option I. It is possible to recommend I without recommending a dominated option.

According to the Gärdenfors-Sahlin approach, refusal is uniquely admissible in all three cases. But in case 3, refusal is not a Bayes solution.

It is, to be sure, possible to object to the criteria of admissibility advanced here as rival to gamma minimax that the choices in the three cases which I have just described violate property α (Sen, 1970, p. 17), otherwise known as independence of irrelevant alternatives in one of its many senses (Levi, 1974, 1980a).

(Let T be a set of options containing x and let S be a subset of T also containing x. A criterion of admissibility defines a "choice function" over subsets of some large set U of options (which we take to include the sets T and S) which takes as values sets of feasible options. Such a choice function has property α if and only if given that $x \in C(T), x \in C(S)$. That is to say, x is admissible in S if it is admissible in T.)

In my judgment, failure to satisfy property α is merely one symptom of the fact that there is no ordering of the options with respect to expected utility due to the indeterminacy of probability judgment. We have to decide whether we are going to abandon the requirement that admissible options be Bayes solutions or abandon property α and with it the demand that the feasible options can be ordered with respect to whether they are better or worse. I favor the latter course. In either case, we have deviated from the path of true Bayesianism - even when Bayesians allow for imprecision in the measurement of probability. The difference between my approach and the approach of advocates of gamma minimax is that they seek an ordering of the options as better or worse alternative to the Bayesian one. I reject that alternative and insist, instead, that when credal indeterminacy is present, there often is no ranking of the options as better or worse. But no matter how one views the matter, the insistence on the presence of credal indeterminacy represents a position substantially different from the strict Bayesian point of view according to which, owing to the frailty of human nature, we are not in a position to recognize numerically definite credal probabilities but we should, nonetheless, proceed as if we were committed to such probabilities as best we can.

Perhaps it will be helpful to those who resist the idea of allowing violations of property α to consider an example where the indeterminacy which leads to violation of property α does not derive from indeterminacy in probability judgment. Consider a choice between three job candidates where the demand is for a good stenographer-typist. Each candidate is given a typing test and a stenography test. The test scores for candidates J, D and L are 100, 91 and 90 for typing and 90, 91 and 100 for stenography. Under these circumstances, the manager may not be able to average the scores into a single score but may be in suspense between different ways to weight the scores. So even though he faces a decision problem under certainty, he is conflicted in his utility judgments and regards many different utility functions as permissible. According to some permissible utility functions J beats D beats L. According to others, L beats D beats J. And there are other weak orderings as well. (For example, L beats D which ranks together with J.) Still no matter what permissible function is used, D is never optimal. So D is not *E*-admissible. But J and L are.

The manager may take the view that when professional considerations cannot decide, he will choose on the basis of the extent to which the candidate belongs to an underprivileged minority. Here D is most underprivileged, J is next and L is least underprivileged.

Thus, the manager chooses J in the three-way choice. Before he can implement the choice, L withdraws. The manager reconsiders and now finds both J and D to be *E*-admissible. That is because J beats D according to some permissible utility functions, D beats J according to others and they are ranked together according to one. This leads him to choose D because D is more underprivileged than J. Property α has been violated.

The breakdown of α in this instance appears in the context of decision making under certainty. The appeal to the gamma minimax criterion utilized by Gärdenfors and Sahlin does not suffice to prevent it. Nor should we want an account of rational choice to prevent it. The office manager's valuations are perfectly sensible. Any theory of rationality which says otherwise displays its own inadequacy.

In this case, the office manager orders the candidates with respect to their competence as typists and as stenographers. In addition, he ranks them with respect to their level of disadvantage. But he cannot legitimately, according to his own lights, aggregate these different valuations into a single "all things considered" valuation which ranks the options he faces as better or worse. The office manager is not obliged as a rational agent to fix on a definite ranking, all things considered. Indeed, sometimes it may be illegitimate for him to do so.

If this is right, we have a precedent for recognizing the occasional legitimacy of similar verdicts in the context of decision making under uncertainty. Counter the approach of Gärdenfors and Sahlin, it may be rational, all things considered, not to rank the options as better or worse due, perhaps, to indeterminacy in probability judgment. We should resist the lust for order when there is no warrant for it.

INDETERMINACY IN UTILITY AND THE ALLAIS PROBLEM

The approach of Gärdenfors and Sahlin is an excellent example of how authors sensitive to the deficiencies in Bayesian dogma are driven to an unjustifiably severe rejection of Bayesianism because of the yearning for order. There are other critics of Bayesianism who, by appealing to other kinds of violations of the sure-thing principle, end up doing essentially the same. The best known criticism of this sort appeals to predicaments which are variants of the problems introduced into discussion by M. Allais in the 1950s (Allais, 1952, 1953). These problems are interesting because they are not examples of decision making under uncertainty. At the same time, they are not exactly examples of decision making under risk. Although numerically definite credal probability judgments can be derived from knowledge of chances, as in decision making under risk, expected utility calculations remain indeterminate due to problems with the utility function.

Suppose a decision maker faces an urn containing 1 red, 89 white and 10 blue balls from which one is to be selected at random. Two options are offered given as follows:

	Red	*White*	*Blue*
E	$1,000,000	$1,000,000	$1,000,000
F	$0	$1,000,000	$5,000,000

Compare the decision the agent would make in this case with the choice he would make were the option as follows:

	Red	*White*	*Blue*
G	$1,000,000	$0	$1,000,000
H	$0	$0	$5,000,000

Observe that in this case the agent can ground his judgments of credal probability in knowledge of objective or statistical probability via direct inference. If there is any indeterminacy in credal probability judgment, it will presumably be due to distrust, on the part of those who are presented with these problems, of the data given them. Still these are explicitly hypothetical problems where the agents are invited to contemplate taking the decisions on the assumption that the objective probabilities are as specified. There should be no excuse for distrust here, and charity suggests that we should not interpret the responses to these problems in this light.

Yet, many authors have confirmed the observation of Allais that the predominant choice in the first problem is E whereas H is the choice in the second problem (Hagen, 1979; Kahneman and Tversky, 1979; and MacCrimmon and Larsson, 1979). If choice reveals preference, E is preferred over F and H over G. And this system of preference violates the sure-thing principle.

But in this case, unlike the Ellsberg problem, we cannot deny that options are ordered with respect to expected utility because of indeterminacy in the probabilities used to calculate expected utility.

If there is any indeterminacy in expected utility here, it will have to be in the utilities assigned to the payoffs - which have been specified in money. And, in the Allais problem, it should not be surprising that there is indeterminacy in these utilities. Even if we agree that the marginal utility of money diminishes with income received and, indeed, that the difference between a million dollars and nothing is greater than the difference between five million and a million, we would still have to determine whether the first difference was more than, less than or exactly equal to 10 times the second difference. It is doubtful whether any one of us is committed to a utility of money function so precise as to allow a univocal answer to this question. But if not, then we may find ourselves in a situation where several distinct utility functions are permissible to use together with our numerically determinate credal state to compute expected utility and where, as a result, both E and F are *E*-admissible in the first problem while G and H are *E*-admissible in the second. Of course, if we withdraw the assumption that marginal utility of income decreases and allow indeterminacy on this point as well, the same result will emerge.

If one evaluates security levels by means of Wald's risk functions, it turns out that assessments of security levels go indeterminate due to the indeterminacy in the utility function. Recall, however, that there is no principle of reason which requires assessing security in Wald's way. That depends on the decision maker's values. And it is entertainable that some agents who do conform to Wald's view of security will revert to an appeal to security levels fixed according to ultimate monetary payoffs when the Wald method fails to yield a verdict.

This attitude towards fixing security levels conforms well to the puzzling behavior in the Allais problem. Since the security level of E is better than that of F, it is chosen in the first problem. In the second problem, the security level is the same for both options; but the second worst outcome is better according to H than according to G, so that H may be chosen. Still E is not preferred to F (i.e., it does not have greater expected utility). And H is not preferred to G. There is no violation of the sure-thing principle.

To avoid misunderstanding, it should be made clear that the suggested analysis of the Allais problem does not recommend that the agent change his utility-of-money function when shifting from the first decision problem to the second as Cyert and Degroot (1974) suggest. In both problems, the same set of utility-of-money functions is permissible and that set of permissible utility functions represents the agent's values in both contexts.

This analysis of the Allais problem is based on the same criteria for admissibility as were invoked in connection with the Ellsberg problem. First a test for *E*-admissibility was deployed. That test identifies the set of feasible options which are *E*-admissible - i.e., which are ranked optimal according to some permissible expected utility function defined over the feasible set. Then lexicographical maximin solutions are determined relative to the agent's assessments of security levels from among the *E*-admissible options. (In some cases, assessments of security levels could go indeterminate. I shall not discuss such cases here; but they are allowed by the proposal.)

The only novelty in the Allais case derives from the fact that the multiplicity of permissible expected utility functions arises from indeterminacy in the utility functions defined over the space of consequences of the feasible options and not from indeterminacy in the probability functions. (Needless to say, indeterminacy in expected utility can be a product of indeterminacy in both utility of consequences and probability.)

We already encountered indeterminacy in utility when discussing the example of the office manager hiring a secretary. In that case, the source of the indeterminacy derived from a conflict in the decision maker's values. The office manager sought a good typist and a good stenographer and when the job applicants available did not combine the desiderata in a smooth way, the office manager faced a conflict in values that could not be resolved prior to making a decision.

When faced with the challenge of "aggregating" several utility indicators representing rival desiderata, the view I am proposing holds that any aggregation representing a potential resolution of the conflict ought to be a weighted average of the utility indicators characterizing the desiderata. I do not pretend to be able to demonstrate the propriety of this approach any more than I can demonstrate the propriety of taking *E*-admissibility as the first criterion of admissibility. But assumptions found in Blackwell and Girshick (1954, pp. 116–119), Fleming (1952, pp. 366–

384) and Harsanyi (1955, pp. 309–321) can be adapted to provide a foundation for this view. I elaborate on this in Levi (1986) and in chapters 8 and 9.

Another way in which indeterminacy in utility arises occurs in cases where agents are paid off in risky prospects and where the expected values of these prospects are indeterminate because of indeterminacy in the credal probabilities. Money might be just such a prospect for most individuals, at least to some extent. Granted that most agents will prefer more money to less, so that all permissible utility-of-money functions will be positive monotonic transformations of cash value, and granted, for the sake of the argument, that the marginal utility of money is decreasing, it is at least entertainable that enough indeterminacy in the utility-of-money function obtains to yield the results widely reported concerning the Allais problem. Indeed, these results can be interpreted as furnishing some indication of indeterminacy in utility-of-money functions.

One could test the conjecture about indeterminacy in another way, thereby obtaining an independent check concerning the extent to which experimental subjects responding to Allais-type problems conform to the model I am suggesting. Offer experimental subjects a lottery with 10 chances out of 11 of winning $5 million and another lottery where the agent wins $1 million for sure. If my conjecture about the Allais problem is correct, the agent should take the second lottery, because both options are *E*-admissible and considerations of security favor $1 million for sure. One might in addition seek to ascertain the greatest lower bound on chances for the $5 million which would induce the agent to choose the first lottery. If this bound is less than certainty (as the approach offered here suggests it should be), then compare the lottery with odds for $5 million slightly less than this greatest lower bound with receiving an amount smaller than $1 million for sure. If there is indeterminacy in utility, it is to be expected that for smaller amounts close to $1 million, the agent will choose the sure thing and if the indeterminacy is considerable, the amount subtracted from $1 million could be substantial. I take it that the obstacles to testing this prediction are no greater than in other experiments concerned with assessments of risk and uncertainty.

Beginning with Allais himself, those who have studied problems like the ones he raised have presupposed that ideally rational agents satisfy certain minimal conditions of consistency which entail, so Allais himself maintained, that the ideally rational agents evaluate lotteries or other such "random prospects" in a manner which induces a weak ordering of these prospects as better or worse. In addition, if an agent faces a choice between two lotteries or "random prospects" (Allais, 1952, p. 38) where the probability of the value of the payoff being less than *m* according to prospect P is no greater than according to prospect P' for all *m* and smaller for at least one *m,* P is to be strictly preferred to P'. (Allais, 1952, pp. 39–41 and p. 78. Allais calls this the "axiom of absolute preference.") P is said to "stochastically dominate" P' (Borch, 1979, pp. 194–195).[4]

It is clear from his discussion that Allais is concerned not merely with the question of describing the risk-taking behavior of humans, but with formulating a conception of the ideals of rational decision making which takes uncertainty and risk into account. Allais quite rightly seeks to put principles of rationality into a framework which avoids ruling out as irrational modes of decision-making conduct which manifest predilections for one sort of goals and values rather than another. He focuses on a thin conception of rationality which addresses issues of consistency alone.

Although Allais' approach to rationality does seem to me eminently sensible in these respects, he betrays his own ambition by insisting that a

4 Suppose the lotteries P and P' utilize the same stochastic process. If, for any given outcome o_i of the process, lottery P pays at least as much as P' and for at least one such outcome pays more, P not only stochastically dominates P' but dominates P' in the ultimate payoff. If P dominates P' in the ultimate payoff, it stochastically dominates P'; but the converse need not hold. If we are given a fixed stochastic process and lottery P yields lottery Q_i for outcome o_i while P' yields Q_i' for outcome o_i, it is possible for P to dominate P' with respect to payoffs in lotteries (so that for every i, Q_i is at least as great as Q_i' and for some i is greater) without dominating P' with respect to the ultimate payoff. We may, therefore, distinguish three senses in which someone can prohibit a lottery from being preferred by another which dominates it

(1) One can prohibit a lottery from being preferred by another which dominates it with respect to some way of specifying payoffs.
(2) One can prohibit a lottery from being preferred by another which dominates it with respect to some privileged way of specifying ultimate or "sure" payoffs.
(3) Given the specification of ultimate or sure payoffs, one can prohibit a lottery from being preferred by another which stochastically dominates it.

Prohibition (1) is the demand insisted upon by Bayesians. It is satisfied by the proposals I have made as well. It can be satisfied even if one refuses to recognize a system of ultimate or "sure" payoffs. Allais and those who have taken his approach seriously begin by identifying a system of ultimate or sure payoffs (often monetary prizes) and regard lotteries or random prospects to be assignments of probabilities to these prizes. Allais is prepared to abandon prohibition (1) and insist on prohibition (3) which, in turn, entails prohibition (2). Prohibition (3) is, in effect, the axiom of absolute preference.

Anyone who, like myself, is skeptical of grounding decision theory on some fundamental system of ultimate payoffs will be dissatisfied with the idea of imposing requirements on preference which presuppose such a system. Although the von Neumann–Morgenstern approach is also concerned with evaluating lotteries in a context where ultimate payoffs are recognized, the requirements imposed can remain satisfied even if the erstwhile ultimate payoffs are replaced by lotteries.

rational agent should be in a position to order his options. Since Allais insisted that the choices of E over F and H over G are eminently rational choices (rightly in my opinion), he concluded that, given this rationality, the choice must reveal preference for E over F and H over G. Consequently, he concluded that rational agents may violate Savage's sure-thing principle or, alternatively, the so-called independence axiom.

But the thesis that rational agents should have valuations of options which induce a weak ordering of the alternatives he appraises is an instance of that dogma which, in my judgment, forms the soft underbelly of Bayesian theory. Just as Ellsberg and Gärdenfors and Sahlin respond to the Ellsberg problem in a manner which retains the weak-ordering assumption while abandoning the sure-thing principle, so too Allais and, following him, writers like E. F. McClennen (1983), misdiagnose the trouble with Bayesianism. They tamper with the sure-thing principle or the independence axiom in order to save the demand that rational agents have valuations satisfying the requirements of a weak ordering.

Those who have attended to Allais' work have often been more interested in exploring the extent to which humans deviate from Bayesian norms of rationality in order to criticize the descriptive, predictive, and explanatory value of Bayesian theory. This seems to be true, for example, of Kahneman and Tversky (1979), MacCrimmon and Larsson (1979), Ole Hagen (1979), Mark Machina (1982), and Chew Soo Hong (1981). However, they all take for granted that the agents' choices in situations like the Allais problem "reveal" preferences which are embeddable in a weak ordering, and seek to devise models for choice behavior which cover the sort of behavior manifested in Allais-type problems while preserving weak ordering. As has become apparent, however, there are several proposals available for achieving this end. More to the point, the debate over the merits of various models often turns on questions like whether behavior according to the models satisfies other conditions like the so-called condition of stochastic dominance. Kahneman and Tversky (1979, pp. 283-284) are prepared to abandon not only the independence postulate, but stochastic dominance.[5] Machina (1982, p. 292), however, objects to their prospect theory precisely because it violates stochastic

5 Prospect theory allows for violations of dominance in ultimate payoffs and not merely violations of stochastic dominance. Kahneman and Tversky seek to mitigate the difficulty by postulating that dominated alternatives are recognized and eliminated by decision makers prior to evaluating random prospects. This leads to the use of a choice function violating property α as Kahneman and Tversky, in effect, point out utilizing an example from Raiffa, 1968, pp. 75-76. Since Kahneman and Tversky seem to take for granted that pairwise choice reveals preference, the implication is that transitivity of preference is violated.

dominance. If the only motivation were descriptive adequacy, there appears to be no basis one way or the other for requiring satisfaction of stochastic dominance. Nor are considerations of explanatory adequacy, simplicity, and the like obviously on Machina's side. Machina's insistence on stochastic dominance looks like prescriptive lawgiving, whether he says so or not.

I do not mean to suggest that there is anything wrong with Machina's insistence on stochastic dominance. I am, however, somewhat puzzled by his devotion to this requirement given that he is willing to abandon the substitution postulate which implies such dominance.

In any case, Machina's efforts to show that one can satisfy stochastic dominance while deviating from the substitution principle, accommodating the Allais phenomena, and preserving ordering can be shown to fail thanks to an argument of Teddy Seidenfeld.[6]

But even if we ignore these substantial troubles plaguing the efforts of those who would follow Allais and MacClennen in seeking a system of norms which demand that agents order their options but allow violation of the independence axiom or the sure-thing principle, there is another consideration which argues against such approaches. None of the schemes for accommodating the responses to Allais' problem which undermine sure-thing principles accommodate the Ellsberg problem as well.[7] However, if one is prepared to abandon the requirement that all alternatives are weakly ordered, one can not only recognize that no violation of sure-thing reasoning is displayed, but analyze both the Ellsberg and Allais problems according to the principles of a single decision theory. When it comes to descriptive adequacy as well as normative attractiveness, the proposals I have been making are superior to any of those which have been advanced up until now - provided one is pre-

6 Consider two lotteries L_1 and L_2 which are both valued at \$10. Machina allows for the violation of the substitution principle through a violation of "mixture dominance" (Chew, 1981, p. 4). In particular a 50–50 mixture of these two lotteries could be valued more than \$10 - say \$25. Substitute the two lotteries for the options L and R in the argument of footnote 3, and the result will be that an option will be recommended which is not only stochastically dominated, but dominated in the ultimate payoffs. As in the argument in footnote 3, the reasoning works only on the assumption that all lotteries are weakly ordered.

7 Chew prohibits violations of the substitution principle through mixture dominance (Chew, 1981, p. 4). The violation of substitution according to Allais can be construed as a case where mixture dominance is satisfied. But the Ellsberg phenomenon is a clear case of violation of mixture dominance. Hence, Chew's proposal cannot, even in principle, rationalize the Ellsberg phenomenon. Whether some variant of approaches like Machina's, which allow for violations of mixture dominance, can rationalize the Ellsberg phenomenon remains an open question; but, as we have seen, such approaches must violate stochastic dominance.

pared to abandon the dogma that alternatives ought to be ordered with respect to value.

There is at least one type of example discussed by Allais (1952, pp. 90–92) and studied by Hagen (1979, pp. 283–297), MacCrimmon and Larsson (1979, pp. 350–359) and Kahneman and Tversky (1979, p. 267) which does not seem to be accounted for by this approach. These examples exemplify the "common ratio effect."

Suppose, for example, that an agent is offered two options with the following payoffs:

I	\$1,000,000 with probability .75	\$0 otherwise.
II	\$5,000,000 with probability .60	\$0 otherwise.

In addition, he is invited to choose between the following pair:

III	\$1,000,000 with probability .05	\$0 otherwise.
IV	\$5,000,000 with probability .04	\$0 otherwise.

Observe that the ratio of the probabilities of \$1 million and of \$5 million are the same in both problems and should, according to an expected utility calculation, yield the same ordering of the options as the substitution axiom requires.

Yet MacCrimmon and Larsson report that out of 17 subjects 6 switched from choosing I to choosing IV, in violation of the independence assumption (1979, fig. 3 (i), p. 357). Similar results are reported by Kahneman and Tversky (1979, problems 7 and 8).

This result cannot be rationalized by the approach I have used before. If there is sufficient indeterminacy in the utility function to render options I and II *E*-admissible in the first problem and III and IV *E*-admissible in the second, the security levels are the same for I and II and the secondary security level for II is superior. This suggests that in the first pair, II ought to be chosen over I – counter to the practice of the experimental subjects.

Of course, it is far from obvious that both options I and II are *E*-admissible in the first problem. When discussing the first Allais problem (the so-called common consequence problem), we concluded that the utility function for the range of monetary values for \$0 to \$5 million rendered it indeterminate as to whether a gamble where one has a chance of 10/11 of winning \$5 million is better, worse, or indifferent to \$1 million for sure – at least for many experimental subjects. However, those same subjects could very well prefer \$1 million for sure to an 80% chance of winning \$5 million. This shows only that the utility of money function is partially indeterminate. And that would account for the choice of I over II.

But then, if the independence postulate is observed, III should be chosen over IV. The experimental data suggest, under this interpretation, that the independence postulate is, indeed, violated.

Observe, however, that the violation can be explained as due to a tendency on the part of experimental subjects not to distinguish probability .04 from probability .05, and to treat them as equal. If they do so, the choice of IV over III becomes the optimal option in expected utility.

Thus, even if it is conceded that in this type of common probability-ratio problem the independence postulate is violated, the violation can be attributed to a tendency on the part of experimental subjects to neglect small differences in probability assignments. In the MacCrimmon and Larsson examples, the differences neglected are in the second decimal place, whereas in the Kahneman and Tversky example (which yielded a more striking shift) the probability differences are in the third decimal place.

It is not my purpose to deny that individuals make errors of this sort and violate widely advocated principles of decision making or that they violate the principles which I myself favor.

But the interest which developed in the Allais common-consequence problem and in the Ellsberg problems derives from a sense that the responses which appear to violate the independence and sure-thing principles are not foolish and cannot be attributed to neglect of the importance of small differences in numerical magnitudes or other errors likely to crop up in rough computation.

It does not seem to be the case that the violation of the independence postulate exhibited in the common ratio of probabilities cases of the sort studied by MacCrimmon and Larsson or by Kahneman and Tversky impose pressure to reconsider norms of rational choice in quite the way the Ellsberg problem or the common-consequence problem of Allais can do. There is a readily available explanation of the failure in terms of the short cuts individuals use in making computations. We may easily concede that the norms of rationality in general and the strong independence principle in particular are violated, and perhaps even that no amount of effort will prevent anyone from failing to observe them on at least some occasions, as long as we can attribute the failures to limitations on computational capacity, the errors we make in our efforts to simplify our computational tasks, or our vulnerability to emotional disturbance. We can, so it seems, admit the existence of violations of the strong independence condition arising in common ratio problems without pressure to modify commitment to the sure-thing principle or strong independence axiom. The experiments done on the common ratio problem exemplify deviations from norms of rationality which point in the direction of educating the experimental subjects rather than modifying the norms.

If this claim is right, then I think we are in a position to say that insofar as the violations of the strong independence axiom identified by Allais cannot be attributed to failures of computation, they are not violations at all but are due to indeterminacy in the agent's utility assessments comparable to the indeterminacy in probability assessments present in the Ellsberg problem and having a parallel impact on the proper analysis of decision problems.

PREFERENCE REVERSAL

This conclusion may be reinforced by considering the following pair of gambles:

1. $1 million with probability .9.
2. $5 million with probability .8.

Suppose the utility function for money is such that U($$x$) is any function in the convex hull of $\log(x+1)$ and $\log(0.1x+1)$.

The expected utilities of the two gambles are 12.44 and 12.34 respectively according to the first function with corresponding dollar values of $252,000 and $230,000. According to the second utility function the expected utilities are 10.36 and 10.49 with corresponding dollar values of $310,000 and $359,500.

Both options are *E*-admissible because they come out optimal according to some permissible utility function in the set. Leximin favors the gamble on the $5 million (gamble 2). It will continue to do so if we modify the gambles slightly. We charge a small amount – say $10 – for the $5 million gamble. Both options should remain *E*-admissible and leximin should continue to favor gamble 2. The range of dollar values for the pair of gambles will be altered only negligibly and it will remain the case that the smallest dollar value for gamble 1 will be higher than the smallest dollar value for gamble 2. Hence, if asked for the smallest price at which he would sell gamble 1, the decision maker should answer $252,000, and for gamble 2 he should answer $230,000. For lower prices, retaining the gamble in question would be uniquely *E*-admissible. For prices of $252,000 and $230,000 respectively, both retaining and selling become *E*-admissible and selling is the maximin solution.

The upshot is that although gamble 2 should be chosen over gamble 1, the lowest price at which the agent ought to sell gamble 1 ought to be higher than the lowest price at which he ought to sell gamble 2. As far as I can see, the situation I have just described reproduces the phenomenon reported by Grether and Plott (1979) and others (Reilly, 1982) on so-called preference reversal.[8] On my account, of course, there is no preference reversal - only indeterminacy in preferences.

NEWCOMB'S PROBLEM

Finally and very briefly, I should mention that the tendency on the part of some respondents to the Newcomb problem to choose the so-called two-box solution is itself explainable as due to indeterminacy in the probability judgments of the agent.

An agent is told that he has a choice between receiving the contents of a transparent box containing $1,000 together with the contents of an opaque box or the contents of the opaque box alone.

The agent is told that a demon is an extremely reliable predictor of his choices. Moreover, the demon places $1 million in the opaque box if he predicts the agent will choose the opaque box alone. Otherwise he places nothing in the opaque box.

When asked how they would choose, it is alleged that people divide between those favoring one box and those favoring two.

One can make sense out of these responses if one keeps in mind that the assumption that the demon is a reliable predictor of the agent's choices would normally be taken to mean that conditional on the demon predicting his choice of one (two) boxes, the probability of his choosing one (two) boxes is high. But, in the standard formulation of the problem, the probabilities of the demon predicting his choice correctly given his choosing one (two) boxes are not stated. To obtain these conditional probabilities from the conditional probabilities given in the assumption

8 I cannot cite decisive evidence that my scheme models the Grether-Plott phenomenon because the reports of the experimental design and results provided by Grether and Plott do not furnish information relevant to this matter. They invite experimental subjects to compare pairs of bets of types 1 and 2 (so-called P-bets and $-bets). Counter to the impression given on p. 623 of Grether and Plott (1979), each bet in a given pair always incurs a risk of loss. However, in some pairs, the losses are greater for P-bets and in some pairs, the losses are greater for $-bets. Grether and Plott do not report, however, the percentages of experimental subjects who choose members of pairs in a way which minimizes such losses (and, hence, maximizes security). Hence, although their results and those of Reilly appear compatible with my model, there is an unsettled empirical question about their experiments which is relevant to the empirical adequacy of the model I have proposed.

of the demon's reliability as a predictor, assumptions need to be made concerning the unconditional probabilities of the demon predicting one way or the other. If the agent's judgments of unconditional probability are indeterminate, the conditional expected utility functions for the two options may go indeterminate. Hence, both options could be *E*-admissible. Maximin recommends the two-box solution (Levi, 1975). This result rationalizes the choice of two boxes without introducing any new principle of causal decision theory but merely by acknowledging the presence of indeterminacy in probability and, perhaps, utility judgment.

CONCLUSION

On the view I favor, the Bayesians are right in giving pride of place to considerations of expected utility. Their critics are right, however, in complaining that some of the ways in which agents appear to deviate from Bayesian ideals of rationality cannot be explained away as due to failures of computational capacity, memory, attention, discrimination and the like, or to emotional upset.

Unfortunately most of the critics of expected utility have failed to come up with prescriptive accounts of rational choice having the generality of Bayesian decision theory. The proposals made to accommodate the Allais problem fail to handle the Ellsberg problem and vice versa. And none seem to handle the preference reversal phenomenon. From the Bayesian point of view, these proposals look like ad hoc repairs. And in my judgment, the Bayesians are right.

This view is reinforced by the fact that all of the proposals made have implications which even their advocates ought to find embarrassing. As we have seen, they tend to recommend the choice of options dominated by other options in a sense in which such choice seems clearly to be avoided.

In this essay, I have argued that the Bayesians are wrong in insisting that probabilities and utilities should be so definite that expected utility can always provide a verdict. Not only does such a view entail the conclusion that the experimental results we have mentioned establish a far more deep-running irrationality in human decision makers than seems necessary to suppose, but it condemns those who are anxious about catastrophic possibilities of various energy policies, of the dissemination of various kinds of drugs, of weapons development policies, etc., to the status of irrational fools.

By admitting that we are in doubt much of the time in our judgments of probability and our evaluations of states and options, we are in a position to make systematic sense of the tension between considerations of expected value and security. The insights of Bayesians and maximiners can be preserved in a principled fashion free from the charge of ad hoc eclecticism which may with some justice be brought against those who attack the priority of the expected utility principle. We have seen that allegedly deviant responses to Ellsberg's problem, the standard Allais problem, the Grether-Plott preference reversal problem and the Newcomb problem can all be rationalized within the framework of a single system of norms for rational choice. The proposals made are more general than the Bayesian theory. Indeed, Bayesian theory becomes a limited special case. The proposals allow a wider variety of behaviors to be nondeviant than either the Bayesian theory or the views of those who attack the priority of the expected utility principle. They do not make fools of us unnecessarily.

As indicated, the scheme advanced here cannot accommodate one kind of deviant behavior connected with problems posed by Allais – to wit, the common ratio problem. But as I have argued, an examination of the cases reported suggests that these deviations may be of the sort which a prescriptive account of rational choice need not be expected to rationalize.

To obtain the benefits claimed for these proposals, the requirement that rational agents be in a position to opt for the best, all things considered, according to some weak ordering of their options as better or worse, has had to be abandoned and in a rather severe manner. The indeterminacies in probabilities, utilities and expected utilities allowed together with the criteria of admissibility proposed undermine widely shared assumptions concerning rationality.

In my view, the benefits clearly outweigh the costs. Even so, more can be done. Extensive experimentation may be undertaken to ascertain the extent to which the prescriptions of the theory proposed here are satisfied by decision makers. And those devoted to order can take up the challenge to construct a theory as general as the one advanced here while still meeting their own requirements. Pending such developments, a system of prescriptions preserving the insights of Bayesians and maximiners ought to be taken very seriously by students of rational choice, especially when it seems to avoid making fools of us all unnecessarily.[9]

9 My first published account of the Allais problem is in Levi, 1982a. In Levi, 1982b, I introduced the analysis of the Ellsberg problem. Two other authors have subsequently offered analyses along similar lines. G. W. Bassett has an unpublished manuscript on the subject. F. Schick (1984, pp. 47–51) blames indeterminacy of probability for the Ellsberg phenomenon and indeterminacy of utility for the Allais problem.

REFERENCES

Allais, M. 1952. "The Foundations of a Positive Theory of Choice Involving Risk and a Criticism of the Postulates and Axioms of the American School," translation of "Fondements d'une Théorie Positive des Choix Comportant un Risque et Critique des Postulats et Axiomes de L'École Américaine." In *Expected Utility Hypotheses and the Allais Paradox,* by M. Allais and O. Hagen, 1979, pp. 27-145. Reidel.

1953. "Le Comportement de l'homme Rationnel Devant le Risque: Critique des Postulats et Axiomes de l'École Américaine," *Econometrica* 21:503-46.

Allais, M., and Hagen, O. 1979. *Expected Utility Hypotheses and the Allais Paradox.* Reidel.

Berger, J. O. 1980. *Statistical Decision Theory.* Springer-Verlag.

Borch, K. 1979. "Utility and Stochastic Dominance." In *Expected Utility Hypotheses and the Allais Paradox,* by M. Allais and O. Hagen, pp. 193-201. Reidel.

Blackwell, D., and Girshick, M. A. 1954. *Theory of Games and Statistical Decisions.* New York: Wiley.

Chew, Soo Hong. 1981. "A Mixture Set of Axiomatization of Weighted Utility Theory." Fourth revision of a 1981 working paper.

Cyert, R. M., and Degroot, M. 1974. "Adaptive Utility." In *Adaptive Economic Models,* edited by R. H. Day and T. Grove. Academic Press. Reprinted in *Expected Utility Hypotheses and the Allais Paradox,* by M. Allais and O. Hagen, pp. 223-41. Reidel.

Ellsberg, D. 1961. "Risk, Ambiguity, and the Savage Axioms." *Quarterly Journal of Economics* 75, 643-69.

Fisher, R. A. 1959. *Statistical Methods and Scientific Inference,* 2nd ed. Hafner.

Fleming, M. 1952. "A Cardinal Concept of Welfare." *Quarterly Journal of Economics* 66: 366-84.

Friedman, M., and Savage, L. J. 1948. "The Utility Analysis of Choices Involving Risk." *Journal of Political Economy* 56: 279-304.

Gärdenfors, P., and Sahlin, N.-E. 1982. "Unreliable Probabilities, Risk Taking and Decision Making." *Synthese* 53: 361-86.

Gibbard, A., and Harper, W. 1978. "Counterfactuals and Two Kinds of Expected Utility." In *Foundations and Applications of Decision Theory, v. 1,* edited by C. A. Hooker, J. J. Leach and E. F. McClennen, pp. 125-62. Reidel.

Grether, D. M., and Plott, C. R. 1979. "Economic Theory of Choice and the Preference Reversal Phenomenon," *The American Economic Review* 69: 623-38.

Hagen, O. 1979. "Towards a Positive Theory of Preferences Under Risk." In *Expected Utility Hypotheses and the Allais Paradox,* by M. Allais and O. Hagen, pp. 271-302. Reidel.

Harsanyi, J. C. 1955. "Cardinal Welfare, Individualistic Ethics, and Interpersonal Comparisons of Utility." *Journal of Political Economy* 63: 309-21.

Kahneman, D., and Tversky, A. 1979. "Prospect Theory: An Analysis of Decision Making under Risk." *Econometrica* 47: 263-91.

Levi, I. 1974. "On Indeterminate Probabilities." *Journal of Philosophy* 71: 391-418.

1975. "Newcomb's Many Problems." *Theory and Decision* 6: 161-75. Reprinted in *Decisions and Revisions,* by I. Levi, pp. 245-56. Cambridge.

1980a. *The Enterprise of Knowledge.* MIT.

1980b. "Induction and Self Correcting According to Peirce." In *Science, Belief and Behaviour: Essays in Honour of R. B. Braithwaite,* edited by D. H. Mellor, pp. 127-40. Cambridge.

1982a. "Conflict and Social Agency." *Journal of Philosophy* 79: 231-47. Reprinted in *Decisions and Revisions,* by I. Levi, pp. 257-70. Cambridge.

1982b. "Ignorance, Probability and Rational Choice." *Synthese* 53: 387-417.

1984. *Decisions and Revisions.* Cambridge.

Luce, R. D., and Raiffa, H. 1958. *Games and Decisions.* Wiley.

Marshall, A. 1920. *Principles of Economics,* 8th edition. Macmillan.

Machina, M. 1982. "Expected Utility Analysis without the Independence Axiom," *Econometrica* 50: 277-323.

MacCrimmon, K. R., and Larsson, S. 1979. "Utility Theory: Axioms versus 'Paradoxes.' " In *Expected Utility Hypotheses and the Allais Paradox,* by M. Allais and O. Hagen, pp. 333-409. Reidel.

McClennen, E. F. 1983. "Sure-Thing Doubts." Draft of paper prepared for the First International Conference on Foundations of Utility and Risk Theory.

Pearson, E. S. 1962. "Some Thoughts on Statistical Inference." Reprinted in *Selected Papers of E. S. Pearson,* pp. 276-83. University of California Press.

Raiffa, H. 1961. "Risk, Ambiguity, and the Savage Axioms: Comment." *Quarterly Journal of Economics* 75: 690-95.

1968. *Decision Analysis: Introductory Lectures on Choices Under Certainty.* Addison Wesley.

Reilly, R. J. 1982. "Preference Reversal: Further Evidence and Some Suggested Modifications in Experimental Design." *American Economic Review* 72: 576-84.

Robbins, H. 1964. "The Empirical Bayes Approach to Statistical Decision Problems." *Annals of Mathematical Statistics* 35: 1-20.

Savage, L. J. 1954. *The Foundations of Statistics.* Wiley.

Schick, F. 1984. *Having Reasons.* Princeton.

Sen, A. K. 1970. *Collective Choice and Social Welfare.* Holden-Day.

Wald, A. 1950. *Statistical Decision Functions.* Wiley.

11
Conflict and inquiry

"Give a dog a bad name and hang him." Human nature has been the dog of the professional moralists and consequences accord with the proverb.[1]

Thus, John Dewey begins his *Human Nature and Conduct.* The theme Dewey means to press is familiar enough in his work. Morality seeks to control human nature which resists regulation and so comes to be regarded as a source of evil to be mastered. But given the intractability of human nature, pressure builds to identify some aspect of human nature which is subject to moral control. Dewey writes: "The severance of morals from human nature ends by driving morals inwards from the public open out-of-doors air and light of day into the obscurities and privacies of an inner life. The significance of the traditional discussion of free will is that it reflects precisely a separation of moral activity from nature and the public life of men."[2]

Dewey sought to draw the contrast between the moral and the non-moral in a way which avoids "the severance of morals from human nature." He granted that morality is concerned with the regulation of human conduct. Perhaps it would be better to say that Dewey granted that moral questions are questions concerning human conduct, for, on his view, the notion of there being a morality consisting of a system of normative principles which are distinctively moral is part and parcel of the severance of morals from human nature he sought to avoid. Dewey wrote: "Conduct as moral may thus be defined as activity called forth and directed by ideas of value or worth where the values concerned are so mutually incompatible as to require reconsideration and selection before an overt action is entered upon."[3]

Thanks are due to Akeel Bilgrami, Charles Larmore, Adrian Piper, Carol Rovane, the participants in the conference on "Pluralism and Ethical Theory" held at Hollins College in early June 1991, and the editors of *Ethics* for helpful criticism.

From *Ethics* 102 (July 1992), pp. 814–34.

1 John Dewey, *Human Nature and Conduct: The Middle Works of John Dewey, 1899–1924,* ed. Jo Ann Boydston (Carbondale: Southern Illinois University Press, 1976), p. 4.

2 Ibid., p. 9.

3 John Dewey and James Tufts, *Ethics: The Middle Works of John Dewey, 1899–1924,* ed. Jo Ann Boydston (Carbondale: Southern Illinois University Press, 1978), p. 194.

As is well known, Dewey was an advocate of bringing scientific approaches not only to the study of the mind and society but to questions of ethics and politics as well. But unlike the utilitarians who fixed on a single standard of intrinsic value and saw the importance of a more scientific approach to human subjects to be the improved efficiency we might achieve in promoting the overall good, Dewey was a value pluralist who saw in the diversity of values which men seek to promote the occasion for conflicts analogous to the difficulties prompting those "real and living doubts" which Charles Peirce had claimed to be the occasion of serious scientific inquiry.[4] For Dewey, the essence of scientific inquiry is intelligently conducted problem-solving activity along the lines of the Peircean belief-doubt model. The resolution of conflicts in values provides one of several types of opportunity for engaging in such activity.

Although Dewey made many gestures toward elaborating a conception of problem-solving inquiry both in science and in addressing questions of value and had worthwhile things to say which, to the detriment of contemporary philosophy, were, for the most part, ignored by postwar North American philosophers, it is, nonetheless, fair to say that Dewey's account of such inquiry remained schematic at the end of his long career and that those who have followed him seem to make it a matter of principle to leave the schematism unsullied by elaboration and modification.

In my judgment, this is unfortunate. The belief-doubt model favored by Peirce and Dewey turns its back on the demand that beliefs and values currently endorsed require justification. According to the Peirce-Dewey approach, we need a justification for changing our point of view whether by ceasing to doubt what was formerly open to question or coming to doubt what was initially taken for granted. Moreover, justifications for changing a point of view ought to show that the proposed change is an acceptable means for promoting the goals of the inquiry being undertaken. When this perspective on inquiry and justification is combined with a rejection of a single fixed standard of value, the key elements of the pragmatism of Peirce and Dewey which I admire are identified. The problematic generated by this version of pragmatism has ramifications of first-rate importance which can be addressed with reasonable intellectual rigor. The conception of value pluralism I mean to explore and defend is a variant of pluralism as understood within the framework of this problematic.

4 C. S. Peirce, "The Fixation of Belief," in *The Writings of C. S. Peirce* (Bloomington: Indiana University Press, 1978), vol. 3.

In the postwar era, Isaiah Berlin has championed a value pluralism which has been seen by him and by his students, for example, Bernard Williams, as furnishing the occasion for value conflict.[5] Yet, in sharp contrast to Dewey, the presence of value conflict is not conceived to be the occasion for inquiry and deliberation, as the belief-doubt model requires. According to Williams, inquiry aimed at removing conflicts in value is inquiry seeking to develop an ethical theory which will iron out value conflicts within the framework of a single coherent system. Such a scheme sees conflict as a form of inconsistency and "applies to moral understanding a model of theoretical rationality and adequacy."[6] Williams rejects this understanding. Value conflict is not necessarily pathological. Indeed, eliminating value conflict typically incurs a loss - "the loss of a sense of loss" - and a flattening of human experience.[7]

Williams stops short of concluding that we should never seek to remove conflicts in values. However, he does appear to take the position that removal of value conflict becomes desirable only when there is some incentive outweighing the costs which he thinks are incurred by doing so. Williams thinks that individuals are capable of tolerating a considerable amount of value conflict in their personal affairs and can manage to make decisions relying on their personal judgment or intuition.

Conflict is not so easily handled in the context of public policy formation. Public agencies ought to be held "answerable" for the policies they promote, and such answerability is accompanied by a need to remove or deny conflict. To the extent that public policy becomes involved in questions of ethical concern, a pressure to rationalize values and remove conflict arises which is at odds with the importance of preserving the multifaceted structure of private values. Williams concludes: "If philosophy is to understand the relations between conflict and rationalization in the modern world, it should look towards an equilibrium - one to be achieved in practice - between private and public."[8]

Thus, even the conflict between the private and public is, according to Williams, not to be seen as the occasion for inquiry but rather for achieving an equilibrium "in practice."

In spite of the recognition of the diversity of value and of conflict in values which Williams and Dewey share in common, they differ pro-

5 See Isaiah Berlin, *Concepts and Categories,* ed. H. Hardy (New York: Penguin, 1978); Bernard Williams, *Problems of the Self* (Cambridge: Cambridge University Press, 1973), *Moral Luck* (Cambridge: Cambridge University Press, 1980), and *Ethics and the Limits of Philosophy* (Cambridge, Mass.: Harvard University Press, 1985).

6 Williams, *Moral Luck,* p. 81.

7 Ibid., pp. 80, 82.

8 Ibid., p. 82.

foundly in their understanding of the way in which such conflict is or ought to be related to inquiry.

According to Dewey, following Peirce, all inquiry, whether it is scientific inquiry or inquiry concerning values, is problem-solving activity. In a given context, some issues are settled and others unsettled. Among the unsettled issues, the inquiring or deliberating agent takes some to be urgent to resolve, if feasible, and so undertakes an inquiry. The conduct of such inquiry is structured, according to Dewey, along the lines of a practical deliberation. The agent has a problem to solve. He needs to become clear as to his goals in solving it and also needs to identify possible means for doing so. Once he has come to an adequate conception of the available means and how likely they are to realize the values which articulate the goals, the agent is in a position to reach a conclusion as to what he should do.

I do not want to place much stock in Dewey's account of practical deliberation as such. It seems to me far too inchoate to be satisfactory for the purpose of a systematic discussion. What is clear, however, is that, for Dewey, the minimal principles regulating intelligently conducted deliberation are applicable not only to economic, political, prudential, and ethical deliberation but to scientific inquiry as well.

To this extent, therefore, there is no difference between "practical" rationality and "theoretical" rationality. The core characteristic of the pragmatism of Dewey worthy of our admiration and serious attention is the thesis that all rationality is in the service of problem solving.

To endorse such a pragmatism, as I do, need not, however, entail a reduction of the aims of scientific inquiry to moral, political, economic, personal, or aesthetic goals. Once one concedes a diversity of values, as Dewey clearly does, there is little pressure to conclude that the cognitive goals and values which ought to be pursued in factual inquiry should be redescribed as moral, political, economic, personal, or aesthetic goals. A concern to avoid error in fixing beliefs and to gratify the demands for information which occasioned the inquiry (including, e.g., furnishing a systematic and simple explanation of some phenomenon) can characterize cognitive values quite distinct from any utilitarian or economic consideration. Any of these cognitive values can, indeed, come into conflict with other noncognitive values the investigator may be interested in promoting and thereby become grist for a moral dilemma.[9]

9 The celebrated prisoners facing the well-known dilemma can recognize the benefits to both of mutual trust in each other's cooperativeness. And in some cases, "wishful thinking" may take over and the agents will come to believe that their partners will cooperate if they do. Such wishful thinking may be justified because it promotes

In spite of this, the reasons which are taken to justify the adoption of one among a number of conjectures as the settled answer to a given question should show that on the available evidence, accepting the erstwhile conjecture into the body of settled conclusions best promotes the cognitive goals - all things considered. The reasons are, in this sense, practical reasons - that is, reasons for adopting one option rather than others in order to realize certain objectives on the basis of the available evidence. The principles (whatever they may be) determining which options are admissible and which are not ought to be the same principles of rational goal attainment that apply when the goals are not cognitive but practical.[10]

Williams is on record as insisting that "there clearly is such a thing as practical reasoning or deliberation, which is not the same as thinking about how things are."[11] One of the earmarks of practical reasoning is its toleration of conflict. In contrast to Dewey, Williams does not see conflict between diverse values as an occasion for reconsidering these values before choice. Williams thinks that efforts to iron out conflicts in value commitments reflect an illicit assimilation of practical reasoning to theoretical reasoning. Just as we seek a coherent systematization of our views concerning how things are, those who indulge in ethical theorizing seek a systematic and coherent account of how we ought to live, what we ought to do, and how things out to be. This theoretical quest for systematicity and coherence in our values tends to reduce our values to a colorless one-dimensional spectrum.

Following Dewey I insist that the distinction between the theoretical and practical is an untenable dualism. Dewey thought that both conflicts between rival value commitments and conflicts between rival conjectures furnish problems for inquiry. The conflict between the null result Michelson obtained with his interferometer experiments at Potsdam when he sought to measure the velocity of the earth in the ether and the predic-

cooperation with the beneficial consequence of reducing the prisoners' jail terms. This is surely a noncognitive value. It recommends forming new beliefs without regard for avoiding error for the explanatory or informational value of the new beliefs. It urges a different policy for fixing belief than a concern to acquire new error-free information which might warrant, in some cases, predicting that one's partner will not cooperate. Often enough Prisoner's Dilemmas and the bargaining situations resembling them are problematic because of uncertainty as to what one's partner will do. In such cases, Prisoner's Dilemmas become conflicts between the cognitive values of avoiding error and obtaining answers and the concern to obtain a Pareto optimum. Because of this conflict between cognitive and noncognitive values, the Prisoner's Dilemma becomes a moral dilemma.

10 Isaac Levi, *Gambling with Truth* (New York: Knopf, 1967), and *Enterprise of Knowledge* (MIT, 1983).

11 Williams, *Ethics and the Limits of Philosophy,* p. 135.

tions of classical theory was untenable and required critical scrutiny both of Michelson's experimental results and of classical theory. Similarly, conflicts between commitments to equality between the sexes and to traditional conceptions of the role of husband and wife in the family call for critical reflection. If anything, it is more difficult to tolerate inconsistency between value commitments than it is between factual claims.

Williams identifies the apparently "first personal" status of practical deliberation as a critical difference between it and factual deliberation.[12] Williams seems to think that the "I" in "What should I think about table top nuclear fusion?" is impersonal whereas the "I" in "Should I comfort my bereaved friend?" is not.

I do not see the difference. Fixing belief involves making decisions just as surely as settling on a course of action does. To be sure, in practical deliberation, the option chosen will not in general bear a truth value. Avoidance of error cannot, therefore, be a desideratum. On the other hand, in factual deliberation, one seeks to add some erstwhile conjecture to the settled assumptions. The conjecture has a truth value. In adding it to the settled assumptions, the inquiring agent may import error into his settled assumptions. Hence, avoidance of error can be a value in factual deliberation. Even so, agents may add conjectures to their settled assumptions without regard for the possibility of error just as Harry Frankfurt's bullshit artist has no concern for truth or falsity in what he says.[13] Peirce objected to fixing beliefs by the methods of tenacity, authority, and a priori reason because they were not sufficiently concerned with avoidance of error.[14] Perhaps we may mark a distinction between factual and practical inquiry by suggesting that one ought to be concerned to avoid error in the former but not in the latter. Whatever the merits of this suggestion, it distinguishes practical from factual deliberation in terms of the kinds of aims these kinds of activity should have or the types of options which are available and not in terms of some contrast between first personal and impersonal considerations.

I have long urged that the aims of scientific inquiry ought to be autonomous in the sense that they ought not to be reduced to other values.[15] However, no two special inquiries have the same goals, so it is

12 Ibid., pp. 66–70.

13 Harry Frankfurt, "On Bullshit," *Raritan* 6 (1986): 81–100.

14 Peirce.

15 Isaac Levi, *Gambling with Truth, Decisions and Revisions* (Cambridge: Cambridge University Press, 1984), pt. 1, and *Hard Choices* (Cambridge: Cambridge University Press, 1986).

misleading, when speaking of cognitive aims, to speak as if there were a single one. What one might say is that the diverse aims of scientific inquiries share in common a concern to obtain new error-free information.

However, it does not seem to me that this claim about the common features of cognitive goals is a principle of factual reason which everyone engaged in factual reasoning ought to endorse.

Some deny that scientific inquiry has or ought to promote autonomous cognitive goals at all. And among those who acknowledge that there are autonomous cognitive goals worthy of being pursued, there is controversy concerning what they are. Peirce and Popper, for example, held to a messianic realism according to which convergence on the true complete story of the world is the ultimate aim of inquiry.[16] Others who, like Dewey, were skeptical of the merits of such a messianic realism, have sought to downgrade or remove a concern to avoid error - that is, falsity - as a cognitive value. Simplicity, predictive power, fruitfulness, explanatory power, familiarity, and other properties attributable to conjectures which are poorly correlated with truth value (except, perhaps, negatively) are made central cognitive values.

My own view has favored a secular rather than a messianic realism, emphasizing concern to avoid error in the problem under investigation rather than in the end of days and seeing the other cognitive virtues as furnishing an inducement to risk error.

I am not concerned to settle these issues here. But I am denying that settling these controversies over the aims of inquiry has any bearing on the common structure of practical and factual rationality. Rationality, as I understand it, is concerned with weak principles prescribing minimal conditions of coherence which propositional attitudes and choices ought to satisfy. The reasons for choosing one option rather than another are relative (i) to the agent's background body of full beliefs as to what is the case (i.e., those assumptions the inquiring agent takes for granted as certainly true) and, relative to that background information, judgments as to what might and might not be the case, (ii) to his judgments as to how likely or unlikely those propositions which may or may not be the case are to be true, and (iii) to the evaluations of these propositions in terms of how desirable it would be for them to be true.

A theory of rationality ought to furnish the following.

16 See Levi, *Gambling with Truth,* ch. 3, and *Decisions and Revisions,* ch. 8. For Peirce's views, see Peirce, *Writings.* For Popper's views, see Karl Popper, *Conjectures and Refutations* (New York: Basic Books, 1966).

i. It should have a characterization of minimal conditions which any potential body of full beliefs ought to satisfy. For example, one ought to be certain of the deductive consequences of the propositions of whose truth one is certain.

ii. It should include a specification of core requirements any system of judgments of probability ought to obey. For example, judgments of probability ought to be representable by a convex set of "permissible" probability functions.

iii. It should contain a characterization of minimal conditions any structure of evaluations of propositions as better or worse ought to meet. For example, judgments of propositions as better or worse ought to be representable by a convex set of "permissible" utility functions.

iv. It should supply an account of how these three factors control assessments of options as admissible or inadmissible for choice in a decision problem. For example, a feasible option is admissible only if it maximizes expected utility relative to some permissible probability and some permissible utility.[17]

An account of these matters would, I take it, be an account of rationality. It would furnish a characterization of rational choice and with that an account of rational (i.e., coherent) belief and valuing. The principles involved would be impersonal in the sense that they would purport to be applicable to all agents. But their universal applicability would be accompanied by a lack of substantive content. They would impose minimal constraints on the propositional attitudes an agent may have at a given time and, in doing so, would also specify constraints on how such an agent should choose among the options available to her or him at that time.

Such constraints on rational choice would not lead to useful results unless one was given the background of full beliefs, probability judgments, and value judgments relative to which a decision is to be reached. Given a full specification of the relevant contextual parameters, one might derive a recommendation as to what one should do which is impersonal and, hence, universalizable as long as the relativities are retained. This is so whether the problem is a practical or a cognitive one. But the context dependence of such judgments is substantial. If one is seeking a form of objectivity in choice which is freedom from context, then objectivity is not to be found in the context of either practical deliberation or scientific inquiry.

17 See Levi, *Enterprise of Knowledge.*

Williams, I suspect, thinks that factual rationality invokes context-independent standards of evaluation. Or, perhaps more accurately, he thinks that the only relevant contextual parameter is the available evidence. In factual inquiry, the aim of the inquiry is settled: it is inquiry after the truth. There is no possibility of conflict about cognitive value – either between cognitive values or, so it seems, between cognitive values and other values.

But even if there were no conflict between the cognitive values in scientific inquiry, scientific inquirers could differ in the evidence available to them. This is a form of context sensitivity. To this extent, factual rationality cannot be impersonal even according to Williams.

I take it, however, that the impersonality of factual deliberation, as Williams sees it, derives from the presence of a single standard of value. He claims this is absent in practical rationality.

But there is no single standard of cognitive value. Value conflicts can arise among those engaged in scientific inquiries because their standards for what counts as simple, explanatorily adequate, non ad hoc, and what questions are worth asking derive from different research programs and, hence, from distinct cognitive aims. Furthermore, the common feature of the diverse aims of scientific inquiries is the quest for new error-free information. The demand for new information is at odds with the concern to avoid error calling for some form of compromise. How trade-offs are made between these desiderata can become an occasion for conflict even when the concern is to render a verdict as to the merits of competing hypotheses on the evidence. So even if we ignore the possibility of conflict between cognitive and noncognitive values, factual inquiry like practical deliberation can be subject to value conflict.[18]

Thus, if Williams were right in maintaining that we should tolerate inconsistent value commitments, I see no reason why this should not be so when it comes to the pursuit of cognitive values as well as other sorts of value. Or if, as I think, inconsistency is to be removed whenever feasible, conflicts between noncognitive values ought to provide as much of an occasion for inquiry as conflicts between cognitive values.

Following Dewey, I would say that when we find ourselves endorsing two value commitments which cannot be jointly satisfied, we ought first to extricate ourselves from inconsistency by modifying our value commitments so that we no longer regard ourselves under an obligation either to perform the act or to refrain from doing so. We should move to a position of suspense where neither performing the act nor refraining

18 See Levi, *Hard Choices,* ch. 3.

from doing so is ruled out as inadmissible. In such a state of suspense, one can if one likes say that one remains in a state of unresolved value conflict. But it is sheer confusion to conflate being in such a state of suspense with regarding oneself as being under an obligation to do something and to refrain from doing it.

I do not want to say that Williams is guilty of such confusion. He seems to be resolutely committed to endorsing the acceptability of being inconsistent in one's values in the sense that one is committed to performing an act and also committed to refraining from doing so as a resting point. Just as Henry Kyburg a long time ago suggested that a person who fully believes that *h* and fully believes that *h'* is not obliged rationally to fully believe that *h* & *h'*, Williams suggests that a person who is obliged to do *A* and obliged to do *A'* is not obliged to do both *A* and *A'*. Williams thinks that by rejecting such a principle for distributing obligation over conjunctions, he can formally save the rationality of the agent committed to what appears to innocent eyes to be an inconsistency in the agent's value commitments. But both Kyburg and Williams achieve, at best, pyrrhic victories.

Consider Kyburg on the lottery paradox.[19] A ticket is to be drawn at random from a thousand tickets. The chance of ticket *i* not winning (*i* ranges from 1 to 1,000) is 0.999. According to Kyburg, the agent should "accept" that ticket *i* will not win for each *i* even though he or she also accepts that at least one of the tickets will win. Kyburg warns against the disease of conjunctivitis which requires an agent who accepts *h* and accepts *g* to accept *h* & *g*. Once cured of the disease, the lottery paradox ceases to be paradoxical for the agent.

However, if by accepting that ticket *i* will not win, X comes to fully believe or be certain that ticket *i* will not win, X comes to fully believe or be certain that ticket *i* will not win, X's full beliefs contain error and, indeed, as long as X has a modest logical intelligence, he will recognize this to be so. Hence, if he cares about avoiding error, as I take it even Williams thinks he should, he should find this situation untenable.

Kyburg could justly protest that my objection presupposes falsely that accepting the proposition that ticket *i* will not win is fully believing it. One might accept a proposition in Kyburg's sense without being maximally certain that it is true.

I grant that prior to accepting a proposition, one might not be certain

19 See Henry E. Kyburg, *Probability and the Logic of Rational Belief* (Middletown, Conn.: Wesleyan University Press, 1961), "Conjunctivitis," in *Induction, Acceptance, and Rational Belief*, ed. Marshall Swain (Dordrecht: Reidel, 1970), and *Logical Foundations of Statistical Inference* (Dordrecht: Reidel, 1974).

that it is true. This is trivial. In genuinely ampliative revision of belief, one changes one's view by adding new information to what is one's body of settled convictions. But if, by accepting a proposition, one means adding a proposition to the body of settled assumptions or full beliefs, then it must be the case that upon acceptance one is committed to full belief that it is true. Hence, if one adds to one's body of settled assumptions all propositions asserting that ticket *i* will not win, the error which ensues is one which could readily be avoided by refusing to add such information.

Kyburg might not understand accepting a proposition to be accepting it into a body of settled assumptions. If not, what sort of acceptance is he talking about? Does X's acceptance have any relevance to how he should or will conduct himself subsequently in either practical deliberation or scientific inquiry? Pending a satisfactory answer to this question, Kyburg's notion of acceptance retains its consistency by being relegated to the status of an epiphenomenon worthy perhaps of a certain sort of aesthetic appreciation unfolded in this or that "intuition" but scarcely to be taken seriously.

Williams's treatment of ethical inconsistency parallels Kyburg's treatment of the lottery paradox. A similar verdict should be reached.

When an agent faces a predicament where he is under an obligation both to perform some act and to refrain from doing so, the standard principles of deontic reasoning suggest that the agent is under an obligation both to perform and not to perform the act. Not so, says Williams.[20] Agamemnon was obliged to sacrifice Iphigeneia and likewise not to do so but was not obliged to do both.

But if the "ought" here is the "ought of deliberation" (i.e., the sense of "ought" in which prescriptions concerning what one ought to do and what it is admissible to do are intended to control the agent's choices all things considered), such recommendations are clearly absurd. As Williams himself would concede, if one ought to do *A* all things considered and one ought to do *B* all things considered, one ought to do *A* and *B* all things considered.[21] Hence, to judge that one ought to do *A* all things considered and that one ought to refrain from doing *A* all things considered implies that all things considered one ought to do *A* and to refrain from doing *A*. This is, indeed, incoherent.

So, perhaps, the "oughts" which do not distribute over conjunctions are moral "oughts" or some other special category of "oughts." Whatever

20 Williams, *Problems of the Self*, ch. 14.
21 Ibid., pp. 184–86.

they may be, they cannot have the relevance to guiding conduct which the ought of deliberation has. As in the case of Kyburg's treatment of the distribution of full belief over conjunctions, Williams's formalistic maneuver leaves us with a conception of obligation eviscerated of all relevance to what purports to be its intended domain of application.

On Dewey's view, Agamemnon was not simultaneously under an obligation to perform two acts which could not both be implemented. It is true that initially Agamemnon embraced two types of value commitment, one of which enjoined loyalty to his countrymen and troops and the other the obligation to protect his daughter's life and safety. Prior to facing the difficulty, we may suppose that Agamemnon took for granted that he would not be confronted with a situation where both obligations were not satisfiable. This turned out to be factually false. But once he recognized this, he could not coherently retain both value commitments unmodified.

How might he modify them? I suggest that we may take a value commitment - whether it is moral, political, cognitive, aesthetic, or personal - to have two components: (*a*) a specification of a domain of applicability and (*b*) a specification of how alternative options, consequences, and conditions in a situation covered by the domain of applicability are to be evaluated as better or worse. Thus, Agamemnon might have been committed by his loyalty to his countrymen to ranking the offering of a human sacrifice over refusing to do so when the former option is deemed necessary to promote the campaign against Troy and also have been committed by his paternal obligation to rank refusing to sacrifice his daughter over sacrificing her. Agamemnon may have been convinced that he would never face a choice where sacrificing his daughter was necessary to promoting the campaign against Troy. But when he revised that belief, it would have been untenable for him to say that it was better to sacrifice his daughter than not and also better to refrain from sacrificing than sacrificing her. It is just as incoherent to rest content with such inconsistent preferences as it is to rest content with believing that the earth moves in the ether and also to believe that it does not.

In the latter case, one should come to doubt that the earth moves in the ether, entertaining as seriously possible that it does and that it does not. In a parallel spirit, Agamemnon might have come to the view that in predicaments like the one he was facing, it is not ruled out as impermissible for him to rank sacrificing his daughter over refusing to do so and likewise it is not ruled out as impermissible for him to rank refusing the sacrifice of his daughter over indulging in the sacrifice. In this sense,

Agamemnon should have come to doubt whether it was best for him to sacrifice his daughter or to refuse doing so.[22]

How does this way of looking at the matter represent an improvement over Williams's way?

In one respect, it represents no advance whatsoever. It does not single out one of Agamemnon's options as the one he ought to do all things considered any more than Williams's account does. But at least one thing is accomplished. By suggesting that Agamemnon should have been in doubt as to what he ought to have done, one avoids allusion to epiphenomenal obligations, moral or otherwise, to which Agamemnon is alleged to be committed under Williams's view. And this accomplishment is not as inconsiderable as it may seem; for it opens up the possibility that Agamemnon might engage in some sort of inquiry to alleviate his ignorance concerning what he ought to do without begging any controversial question and without remaining in an inconsistent or incoherent state. Agamemnon's problem could then be seen as a moral problem in the only sense in which Dewey thought moral problems are worth considering. And the urgency of engaging in an intelligently conducted inquiry to solve the problem is precisely the urgency of settling on some conclusion as to what ought to be done all things considered. By condoning and,

22 Walter Sinnott-Armstrong, in *Moral Dilemmas* (New York: Blackwell, 1988), pp. 176–77, contends that this account of moral conflict and dilemma is acceptable when one is confronted with competing overriding moral requirements. However, he contends that in moral dilemmas, one is faced with competing nonoverriding moral requirements. "A moral reason to adopt an alternative is a moral requirement if there were no moral justification not to accept it" (p. 12). A moral requirement is not overridden if there is no other moral reason which is more important than it. Thus if a moral requirement recommending the choice of option A were the sole consideration in a given context of choice, one ought, all things considered, to do A. If there are several relevant requirements in the context of choice including the requirement to do A, the given moral requirement is overridden if one is not obliged to do A all things considered. I conjecture that Sinnott-Armstrong's use of the distinction between overridden and nonoverridden requirements is too coarse grained. A requirement may be overridden in the strong sense that not only is it not the case that one ought to do what it requires all things considered but also one is not permitted to do it all things considered. A requirement may be overridden in the weak sense if it is not the case that one ought to do what it requires all things considered whether or not one is permitted to do what it requires. When an agent finds himself in a predicament where two value commitments apply but cannot both be satisfied, the agent may be said to face two competing requirements where each overrides the other in the strong sense. This is a case of inconsistency. Sinnott-Armstrong seems to agree. When the agent revises his value commitments and, hence, moves to a position of suspense, the two requirements are overridden in the weak sense but not in the strong sense. So my account of dilemmas agrees with Sinnott-Armstrong's condition that competing requirements not be overridden provided that this is understood in the strong sense.

indeed, advocating inconsistency, Williams cuts off the possibility of inquiry.

Williams would insist no doubt that only his view does justice to our moral psychology. No matter what the outcome of inquiry of the sort envisaged by Dewey might be, Williams would insist that Agamemnon must do wrong by betraying either his countrymen or his daughter. As is well known, his evidence for this is that no matter what he did, Agamemnon should have displayed regret or remorse for having done it.[23]

This evidence fails to sustain Williams's point.[24] Agamemnon is, as we have said, committed prior to the tragic predicament to an obligation to promote the campaign against the Trojans and to protect his daughter. At that time, the commitment to promote the campaign would have constrained him to prefer undertaking what is required to get the ships sailing rather than to keep them stilled. But if the options are (*a*) to get the ships sailing, (*b*) to keep them stilled but with some apology to the troops, and (*c*) to keep them stilled with no apology, not only is *a* ranked over the other two but *b* is ranked over *c*. A similar remark applies to his commitment to protect his daughter's life. If the options are (*a'*) refusing to sacrifice her, (*b'*) sacrificing her but with suitable remorse, and (*c'*) sacrificing her but without the appropriate grief and remorse, *a'* is ranked over *b'* and *b'* over *c'*. As Agamemnon discovers to his distress, he is faced with a situation where his options are *a* & *c'*, *a* & *b'*, *a'* & *c*, and *a'* & *b*. In doubt, he regards the ranking of *a* & *b'* over *a* & *c'* over *a'* & *b* over *a'* & *c* as permissible and also the ranking of *a'* & *b* over *a'* & *c* over *a* & *b'* over *a* & *c'* as permissible.

Under the circumstances, two options are admissible. One, *a* & *b'*, involves sacrificing Iphigeneia with grief and setting sail whereas *a'* & *b* entails refusing to sacrifice Iphigeneia with apologies to the troops – perhaps accompanied by resignation of his captaincy.

Thus, Agamemnon remains in doubt as to what he ought to do, but he is clear that whatever he does he ought to make amends for what he does not do. This account does not require saying that Agamemnon remains under an obligation to sacrifice his daughter and under an obligation not to with all the mystification that entails. As far as I can see, it accommodates the insight supported by Williams's evidence but it does so without requiring us to say that what the evidence shows is that Agamemnon faced a situation where he does wrong no matter how

23 Williams, *Problems of the Self*, ch. 14.
24 Levi, *Hard Choices*, ch. 2.

he chooses.[25] Rather, Agamemnon confronts a predicament where it is doubtful or unclear or indeterminate as to what he should do all things considered.

Such a predicament constitutes an example of a moral problem in the sense of Dewey - not because moral principles are in question in some special sense but rather because there is doubt as to what one ought to do of the sort which calls for inquiry.

Williams and others like Charles Larmore or Martha Nussbaum who seek to emphasize the moral significance of tragedy will no doubt complain that understanding Agamemnon's predicament in the manner just sketched neglects the futility of inquiry in such cases.[26]

There are two relevant ways to understand such futility. Larmore quite explicitly acknowledges the propriety of suspending judgment in the context of value conflict "when we have some reason to believe that there may come further information that will alter our perception of the situation."[27] Larmore continues: "But not every conflict of this sort arises out of ignorance. Sometimes we are rightly confident that no new decisive facts are in the offing. We know that the conflict is irresolvable. Then, far from knowing too little, we know too much."[28]

Larmore's view represents a distinct advance on the views of Williams and other epigones of Isaiah Berlin who have a penchant for making a mystery out of value pluralism; for he acknowledges the possibility of an alternative way of thinking about conflict even if he thinks it applies only on some occasions - that is, he acknowledges that sometimes there is a point to suspending judgment as to what ought to be done all things considered - namely, when there is hope of resolving conflict through inquiry.

Recall that I claim two benefits from characterizing the appropriate response to value conflict to be to move to a state of suspense: (1) It

25 To be sure, Agamemnon may find himself in a situation where it is not feasible to make amends so that *b* and *b'* are not optional. Yet, so it may be argued, Agamemnon should still feel or acknowledge regret for, say, sacrificing his daughter. I grant the point, but this does not mean that he must acknowledge that he has "done wrong" all things considered. We judge Agamemnon should feel regret because doing so is the mark of a good character, of a person who will make amends if the opportunity were to arise. As I have argued, making amends does not entail an acknowledgment that one has done wrong. Neither does acknowledging regret.

26 See Charles Larmore, *Patterns of Moral Complexity* (Cambridge: Cambridge University Press, 1987); and Martha Nussbaum, *The Fragility of Goodness* (Cambridge: Cambridge University Press, 1986).

27 Larmore, p. 149.

28 Ibid.

avoids insisting that obligations remain in force which either obfuscates or, if not, entails, as it does for Williams, rendering the obligations irrelevant for conduct. (2) It allows for the possibility of engaging in inquiry into the values in conflict without begging questions.

Larmore's complaint is that the second benefit is no benefit at all in cases where there is no hope of resolving a conflict.

Let us suppose Larmore were, counter to fact I believe, right about this. From this it does not follow that when inquiry is pointless one should not adopt a position of suspense but should, as Larmore apparently wants to do, regard both obligations as remaining in force. Suspense remains a preferable posture for the first of the reasons I marshalled - to wit, because obscurantism is eliminated without rendering obligation irrelevant to conduct. As it turns out, Larmore consigns such obligations to irrelevancy when he insists that on those occasions when there is no hope of resolving conflict "ought" no longer presupposes "can."[29] Agamemnon ought not to have sacrificed Iphigeneia and when faced with the decision, perhaps, recognized this. And likewise he ought to have sacrificed her and understood this as well. Unlike Williams, however, Larmore seems to think that standard principles of deontic reasoning remain in place and Agamemnon ought both to have sacrificed Iphigeneia and refrained from doing so. But, Larmore contends, "ought" does not always presuppose "can" - in the sense of feasibility. To my way of thinking, this constitutes as egregious an abdication of the connection between moral reasoning and conduct as anything to be found in either Kant or Williams. Larmore puts the best face he can on this abject concession to irrelevance in the following passage:

> Our deepest moral commitments - that we ought to abide by the strictest deontological requirements and that we ought to bring about the greatest urgent good overall - are commitments whose meaning for us (whatever their origin) is that we *come* with them *to* the world, and not that we *infer* them *from* the world. That is, their role is not even that of (scientific) "background knowledge," something we once learned, which guides our inquiries now, but which in principle remains subject to revision or rejection. Instead, these commitments are what make us moral agents at all.[30]

Perhaps Agamemnon's predicament might be recast as a conflict between promoting the greatest good overall (by sacrificing Iphigeneia) and strict deontological requirements (by not sacrificing her). But this need not be so. What may be called into doubt is that sacrificing Iphigeneia

29 Ibid.
30 Ibid., pp. 149-50.

promotes the greatest overall good or that categorical imperatives prohibit sacrificing Iphigeneia.

But, if, as Larmore seems to think, the tension is between the so-called consequentialist principle and the deontological one, I cannot see how it helps us to be told that then Agamemnon should not suspend judgment because he would violate the requirements for being a moral agent. As we have seen, this means that in that case being a moral agent has no relevance to Agamemnon's conduct. If there is any grand tragedy here, the tragedy is for conceptions of morality which lead to this result - the result to which Dewey pointed when he complained of the "separation of moral activity from the nature and the public life of men."

Larmore himself rightly acknowledges earlier in the same book from which this discussion is taken that value commitments in general have two components: a specification of scope of applicability and a constraint on how options in a problem falling within the scope are to be evaluated.[31] He contends in his first chapter that in general moral principles do not and cannot afford us complete specifications of the scopes of applicability or the constraints involved. To fill the slack, Larmore appeals to Aristotle's conception of *phronesis* or judgment.

Even putting aside the daunting task of clarifying the notion of *phronesis,* there is a confusion in his understanding of the service *phronesis* is intended to render relevant to our current concerns. If the scope of applicability of a principle is not clear and an agent faces a situation which is in the gray area, then it is an open question as to whether the constraint specified by the principle applies or not. We have here a good example of unresolved conflict due to the fact that although the applicability of the constraint to the options faced in the predicament is not ruled out, it is not mandated either. The conflict is not the product of being compelled to doubt what we initially did not doubt due to unexpected conflict. The issue had been an open one all along, but the urgency to settle it had not arisen until the agent faced the situation falling in the gray area. For Dewey, and for me, this too is an occasion for moral inquiry if the question is an urgent one. Appeal to *phronesis* is no substitute for inquiry unless it is an alternative tag for inquiry involving the exercise of imagination required at what Peirce would call the abductive phase (i.e., in proposing potential answers to a question) and the explorations of ramifications of the potential answers obtained which would be required in order to assess their relative merits.

Thus, even if Larmore were right in insisting that a deontological and a

31 Ibid., ch. 1.

consequentialist principle are both constitutive of our conception of moral agency, all that this might mean is that these principles have nontrivial scopes of application while leaving entirely open what these scopes are. Such a view is consistent with acknowledging that there are some situations where only one applies or neither one applies and also that there are plenty of cases where it is unsettled as to whether one or the other applies.

Larmore's appeal to the futility of inquiry maneuver is an analogue to the view often espoused by hyperrealists according to which there may be truths of theory or fact which are inaccessible to inquiry. I myself am sufficiently sympathetic to Peirce's injunction against placing roadblocks in the path of inquiry to insist on taking such roadblocks seriously only when I am given an impossibility theorem. If not, I follow Peirce in refusing to allow for incognizables. And, following Dewey, I see no reason to take a different view when it comes to values.

But let us suppose that we should adopt Larmore's view (which in these matters seems to mimic the ideas of Berlin, Williams, Nagel, et al.). I fail to see why, when faced with incognizables, we should do anything other than we would do were there a hope of successful inquiry – to wit, admit our ignorance and remain in suspense pending the outcome of such inquiry. Perhaps we shall never know the detailed structure of black holes or the precise nature of the processes immediately pursuant to the big bang. We may speculate about the matter all we wish as long as we heed Isaac Newton's injunction never to confuse conjectures with settled certainties.

If this is so in science, why should it be different in ethics? Perhaps, tragic choice is choice in situations where we not only do not know but in some sense could not know what ought to be done. To admit this, if we must admit it, is not to concede that in tragic choice we are under inconsistent obligations. Acknowledging our ignorance is a mark of wisdom. Confusing it with contradiction is just a mistake.

The most obvious and serious answer to this argument is that making a choice is unavoidably peremptory. Sooner or later we must make up our minds or forfeit our autonomy. To guide our action, we must have a view of what to do by the moment of choice. Suspension of judgment is a luxury we can ill afford.

This objection, however, applies not only to cases where conflicts between competing value commitments are irresolvable in principle but also to cases where resolution of conflicts cannot be achieved before the moment when the choice has to be made or where such resolution entails excessive costs. The lack of opportunity to resolve conflicts through inquiry is pervasive in our deliberations even on the most mundane practical matters. If the irresolvability in principle of a value conflict does not by itself prohibit us from suspending judgment, one cannot say that the need to make decisions in such tragic predicaments does so. If it did, it would appear to prohibit suspense in less tragic predicaments as well. One cannot distinguish tragic choices from these more mundane matters by appeal to the peremptoriness of choice since the peremptoriness is often forced by the contingencies of specific circumstances.

In any case, if conflict goes unresolved by the moment of choice so that we cannot determine what ought to be done all things considered, that will be so whether we say that the competing obligations remain in force or we say that we are in doubt as to what we ought to do. Unless we take inconsistency to be the spice of life, clarity requires that we admit to ignorance and that, in many cases, our value commitments are not strong enough to guide our conduct.

Once we admit this much, we may be prepared to concede that rationality does not require that the choices we must perforce make at the moment must be for the best. In a state of ignorance concerning what one ought to do, there may be no best option. The best we may be able to say is that we should avoid choosing any option whose optimality is precluded by all standards of value which remain potential resolutions of our state of doubt and conflict. Given that we must choose, we are rationally entitled to choose any option which is admissible - that is, such choice is not ruled out in this way. Had Agamemnon been wise, he would have conceded that he did not know whether his decision to sacrifice his daughter was for the best. At best he could say that it was admissible - it was not ruled out.

Someone may, perhaps, suggest that at moments like that the deliberating agent needs intuition, *phronesis,* or judgment to carry him through. But *phronesis* amounts to nothing more than a rather obscure way of rationalizing the choice made as for the best once it is made. It is a deus ex machina for resolving conflicts at the moment of choice when reasons justifying the choice relative to the background of belief and value commitment at that moment cannot be provided.

I prefer an attitude which admits that when conflicts are unresolved by the moment of choice, no admissible option is justified as compared to any other admissible option. To the extent that this predicament seems untenable, it ought to provide the occasion for further reflection even after the moment of choice. Lessons may be learned from the chain of events initiated by the tragic decision which can help us address other predicaments better prepared. Far from absolving us from the need to engage in problem-solving inquiry, tragedy ought to prompt us to recognize how urgent it often is.

There remains, however, an important dimension of value conflict as raised by Williams to be considered – to wit, the question of an equilibrium between the private and the public.

Agamemnon was the leader of his sturdy band. His decision was a public one, carrying with it a certain answerability to his troops. It is notorious that leaders who concede that they have no basis for claiming that the choice they make is for the best soon find their hold on the helm slipping from their grasp. Whether it is even in principle feasible to educate a public to tolerate admissions of ignorance as to what ought to be done from their leaders and administrators when such admissions are warranted is at best an open question.

If public officials are answerable for their policies (something we may reasonably demand), such officials will tend to proceed as if their policies are, indeed, for the best all things considered. When there is no consensus as to what is for the best and no consensus that a public policy should be adopted, it would afford an argument (not necessarily a decisive one) for relegating the issues to the private sector. In this way, the prospect of arbitrarily resolving an unsettled issue would be minimized.

This consideration seems to be a better basis for drawing a distinction between the public and the private than the desirability of preserving the multidimensional aspect of our moral experience as Williams suggests.

We are now in a better position to appreciate the difference between the value pluralism suggested by Dewey (or, at least, that version of Dewey's pluralism I have been pressing) and the value pluralism recognized by Berlin, Williams, Larmore, Nussbaum, et al.

For these authors, the plurality of values precludes systematization of our value commitments. The several dimensions of value are irreducible to one another and cannot be incorporated into a single ethical theory. Instrumentalists concerned about scientific theories are fond of insisting on a plurality of rival scientific theories, all of them being applied without any concern for the inconsistencies that threaten to arise. Indeed, the conflict is sometimes seen as a positive benefit which it would be unfortunate to eliminate. The Berlin-Williams value pluralists are the instrumentalists of ethical theory insisting on preserving and fostering ethical inconsistency. Williams explicitly represents his view as opposed to "ethical realism."[32]

Ironically enough, Dewey is regarded as an instrumentalist when it

32 Williams, *Problems of the Self*, pp. 204–5.

comes to scientific theories. To the extent that this is so, I do not share his views. But Dewey's value pluralism is not instrumentalist in the way the Berlin-Williams view is. For Dewey, when we recognize conflicts between our value commitments, inconsistency is not to be tolerated.

The issue, it should be noticed, is not whether value commitments which are the constituents of ethical "theories" bear truth values like the hypotheses in scientific theories. Value commitments do have truth value bearing presuppositions - namely, that the constraints on how options are evaluated as better and worse can be satisfied for decision problems within the scope of the commitments. But the constraints themselves are neither true nor false. The point parallels a similar issue pertaining to probability judgment. An agent may endorse certain conditions on how hypotheses are to be assigned credal or belief probabilities in a given evidential situation. His commitments will be inconsistent if they cannot all be satisfied - if, for example, he is constrained to assign some hypothesis *h* the credal probability 0.4 and also the credal probability 0.6. To acknowledge that such inconsistency is rational is not to endorse the idea that judgments of credal probability bear truth values or that constraints on such judgments are true or false. I am an antirealist concerning both judgments of probability and value commitments insofar as antirealism is the view that such judgments lack truth values. But I stand with the realists when it comes to the desirability of avoiding inconsistency in both probability and value judgment.

When faced with inconsistency in value commitment, we should move to a position of suspense - of acknowledged ignorance. When this happens, our ethical systems become more incomplete than they were before. Certain issues which were initially taken to be settled become unsettled. It could happen that as a result of subsequent inquiry, certain dimensions of value might be integrated into a system which qualifies as reduction to a single dimension. But then again, this need not happen. Scientific inquiry has not always resulted in reductions - however these are to be understood. But inconsistency is tolerated only as long as necessary and, in any case, is not confused with ignorance. The same should hold mutatis mutandis in the case of values.

So for Dewey, perhaps, we should say that value pluralism is better understood as an acknowledgment of our ignorance as to how we ought to live. In ethics as in science, we know some things and do not know others. Our inquiries seek to remove some dark corner of ignorance and to replace it with the light of knowledge, but our efforts often generate new difficulties and cast into doubt erstwhile settled assumptions. Ethical theorizing aimed at systematizing and understanding the blooming buzzing confusion is encouraged rather than prohibited. Attempts to reduce several dimensions of value to a single one are not nipped in the bud. On the other hand, efforts at systematizing our values have not resulted in and are not likely to result in a system of value commitments so complete as to regulate in a determinate way every aspect of our lives. Nor are they likely to issue in a system of ultimate values immune to revision which will form the core of any ethical system - unless such a system is so weak and uninformative as to be useless. We should no more expect progress toward a complete ethical understanding at The End of Days than we should expect progress in science toward a true complete knowledge of the world. Such messianic dreams ought not to control our secular lives. At the same time, we should not grovel in the condition of partial confusion and partial enlightenment in which we find ourselves or, making a virtue out of necessity, declare how dull and flat our lives would be if per impossible we survived Armageddon and faced the Messiah at The End of Days. That way urges us to be complacent in our confusion rather than serious about seeking to remove it. While not dreaming the messianic dream, we should not impose any limits in advance of inquiry on the extent to which we may succeed in straightening out the conflicts in value which we face.

12
The ethics of controversy

Toleration, according to Isaiah Berlin (1969, p. 184), "implies a certain disrespect. I tolerate your absurd beliefs and your foolish acts, though I know them to be absurd and foolish". This view, which Berlin attributes to John Stuart Mill, implies that "sceptical respect for the opinions of our opponents" is "preferable to indifference or cynicism". But even these attitudes "are less harmful than intolerance or an imposed orthodoxy which kills rational discussion" (1969, p. 184).

Berlin (1953, pp. 3–4) claimed that Tolstoy "was by nature a fox but believed in being a hedgehog" and Mill, though declaring loyalty to utilitarian doctrine, was in rebellion against the views of his father and Bentham. Attributing views to others contrary to those they explicitly endorse is a risky business. Berlin's construal of Mill's ideas on liberty and toleration should perhaps be treated with a respectful and cautious scepticism.

We do not have to decide, however, whether Mill endorsed the views on toleration Berlin attributes to him in order to explore the issues raised by Berlin's remarks concerning toleration and respect for others and their views.

In the remarks quoted in part above, Berlin insists that toleration of the views of others does not require a detached attitude towards the issues under dispute or avoidance of firm, deeply and passionately held views on these matters. Precisely because one may hold views deeply and passionately, one may reveal a disrespect for the views of others. In elaborating on his reading of Mill, Berlin writes:

> He believed that to hold an opinion deeply is to throw our feelings into it. He once declared that when we deeply care, we must dislike those who hold the opposite views. He preferred this to cold temperaments and opinions. He asked not necessarily respect for the views of others - very far from it - only try to understand and tolerate them; only tolerate; disapprove, think ill of, if need be mock or despise, but tolerate; for without conviction there are no ends of life,

Thanks are due to Akeel Bilgrami for help in opening up my mind.

and then the awful abyss on the edge of which he had himself once stood could yawn before us. (1969, p. 184)

Berlin's insistence that deep conviction and passion are compatible with rational inquiry and civilized discourse merits our admiration. But his remarks concerning the relation between deep conviction and passion are misleading and may tend to obscure our understanding of the abyss that absence of conviction threatens to foist on us.

I am utterly convinced that $2 + 2 = 4$, that I am not a product of a virgin birth having both a mother and a father and that the laws of Newtonian mechanics apply to an excellent albeit imperfect degree of approximation to middle sized objects moving at moderate velocities relative to the earth. Yet, I hold these deep convictions with no great passion or intensity. I often find myself becoming intense about issues concerning which I lack a sure opinion - although I do not claim perfect correlation in this direction either. I may be eccentric in this respect but I doubt it. It does not matter. My point is that depth of conviction ought not to be equated with either the intensity of the feeling of conviction, the liveliness and vivacity of the associated ideas or the strength of the disposition to assent upon interrogation.

Deeply held convictions concerning what is true and what is false constitute the standard whereby an inquirer evaluates truth value bearing hypotheses with respect to serious possibility. As such, they are commitments or undertakings to judge propositions to be seriously possible if and only if they are consistent with one's explicitly recognized deeply held convictions. As such, these deeply held convictions serve as a resource for guiding practical deliberation and scientific inquiry.

Thus, someone who told me that I was born of a virgin would not arouse any passionate or intense expression of disdain. I might indulge in some vulgar humor; but if I were in a mood to be considerate, I would try to remove the smirk from my face and avoid giving offense. Yet, I would remain absolutely certain that what I was told is false. I would reject the bare logical possibility that it is true as not being a serious possibility.

I would tolerate the opinions of this harbinger of glad tidings. I would not try to stifle the glad tidings or incarcerate its messenger. As Berlin rightly notes, however, my toleration would be tinged with a certain disrespect.

The disrespect would not derive from any passion. To the contrary, it would signal the contempt (and, perhaps, the pity) with which the views expressed by the messenger would be dismissed without consideration of any serious kind - no matter how polite the dismissal might be.

Berlin also mentions that one might adopt a posture of "sceptical respect for the opinions of our opponents". I am not entirely clear whether Berlin intends to distinguish such sceptical respect from the sort of contemptuous toleration described above and, if he does, how he intends to do it. In any case, I think he points to a distinction well worth ferreting out.

I have contempt for the view that I am born of a virgin even though I tolerate its advocates and their expression of it and even though my contempt is unaccompanied by any great intensity of feeling. I am certain that I was not born of a virgin. There is no serious possibility that I was. In what sense, if any, does sceptical respect for the hypothesis of virgin birth differ from this attitude?

Berlin might be interested in studying the manner in which I treat the dissenter from my views. If I am civil and of patient temperament, I may show the dissenter one sort of respect by feigning to take his views seriously in the course of seeking to persuade him of his error.

To show him that I am not the product of a virgin birth by means of arguments which avoid begging the question, I shall have to feign an open mind as to whether I am a product of a virgin birth. I shall pretend without any sincerity retaining all the while my conviction that the claim is preposterous, that there is no serious possibility that it is true. My contempt will be masked by my willingness to pay attention. According to the scenario I am envisaging, I do so not out of respect for that view but, perhaps, because of good manners, kindness or some moral conviction that it is our duty to teach the benighted among us the error of their ways.

Such pseudo respect differs from contemptuous toleration only in its overt expression. The underlying attitude towards the views of the dissenter remains much the same in both cases.

Genuine respect calls for a different response. Instead of feigning an open mind, I must genuinely entertain the new view as a serious possibility. I must change my convictions. Instead of persisting in the conviction that I am not the product of a virgin birth, I must come to suspend judgment as to the truth of that proposition. I must begin to regard both the hypothesis of my virgin birth and its negation as serious possibilities and prepare myself to explore the merits of the rival alternatives without question begging prejudice in favor of one side or the other.

Proceeding in this manner shows respect for the dissenting view not by becoming converted to it but by opening one's mind to the possibility that it is true. This sort of respect is not a species of toleration of views with which one disagrees. To disagree is to be convinced that the rival view is false. Even if one discusses the other view feigning an open mind with the hope of showing the benighted the error of their ways, such disagreement remains contemptuous. Such contempt is not at all like the sceptical respect just characterized. Sceptical respect is manifested only when one is willing to open one's mind up to the new view and engage in a serious effort to examine its merits. That involves a change of mind, an alteration of viewpoint, a willingness to suspend judgment where one was originally in no doubt at all.

Berlin appears to think that in a liberal society of the sort envisaged by Mill (or at least the version of Mill Berlin admires), it is better to open up one's mind to dissent than to refuse to consider it; but if one's mind is closed, tolerating the views of the benighted is preferable to their suppression.

Insofar as Berlin does rate an open mind over a closed one, it is unclear whether Berlin aims to gloss Emily Post on the etiquette of conversation or whether he is concerned with the cognitive attitudes we should take toward dissenting views. Perhaps, he thinks it is better to pretend that one's mind is open to dissent than to cut off conversation without genuinely favoring an open mind over contemptuous toleration. Perhaps, Berlin doubts the viability of the distinction between the etiquette of conversation and the ethics of controversy. If so, he should explain why manners make the man.

I suspect that good manners ought to have some relation to our ideals. There are many occasions where we consider it important to communicate our convictions in as convincing a way to others as possible. Effective communication does often, as a matter of fact, require catering to the sensibilities of those we seek to persuade. Insofar as conforming to the etiquette of conversation is an effective means of persuasion, it may, perhaps, be thought to be of paramount importance.

But persuasion is not everything. Honesty matters. Catering to the sensibilities of others may lead to the substitution of arguments one judges persuasive though bad for good but less compelling reasons. And conforming to an etiquette of conversation can often stifle both imagination and thinking. What is initially advertised as an instrument for effective communication of clearly and sincerely held views often promotes obscurantism and deceit.

I do not want to discount the etiquette regulating the communicative process altogether. But the excessive importance attached by many to communicative procedure – even to the extent of identifying rational conduct with following such procedure – can be destructive of thought and genuine inquiry. To show proper appreciation of the etiquette of debate and conversation, one ought to pay attention to the ethics of controversy – i.e., to those values which should guide the way we form our convictions in the face of disagreement.

I wish, therefore, to discuss that interpretation of Berlin's interpretation of Mill's liberalism according to which it is better to open one's mind to dissent than to refuse it serious consideration but, if one refuses serious consideration, it remains better to tolerate dissent than to suppress it. I deny that this ranking of alternatives is universally applicable. No single ranking of the three alternatives is going to be appropriate on all occasions where such a ranking may be required. Sometimes Berlin's assessment will prove acceptable. On others, an alternative ranking will be appropriate; and sometimes it may turn out that an unresolved conflict between rankings arises.

Philosophers tend to treat the question of coming to believe a proposition and the question of ceasing to believe a proposition with an unwarranted asymmetry.

When an inquirer contemplates adding a hypothesis to the stock of assumptions to be taken for granted (for use as a "resource for deliberation" as John Dewey felicitously puts it), there is broad consensus among philosophers that the inquirer should be in a position to justify the change in view at least to him- or herself.

It is rarely acknowledged, however, that opening up one's mind by removing an item from the stock of settled assumptions and treating an erstwhile item of evidence or settled doctrine as a mere hypothesis calls for justification. On this matter, most writers tend to endorse one or the other of two polar opposite positions: According to one view, nothing could justify opening up one's mind once one has ruled out the possibility that the thesis taken for granted is false. According to the second view, opening up one's mind is warranted each and every time there is any conflict between one's own view and the views of others.

There is, to be sure, a third widely held opinion about these matters. Speaking of justification, whether of coming or ceasing to believe, runs counter to an antivoluntarist orientation deriving from Hume and favored by Quine according to which we lack control over our beliefs. Quinean dogma insists that no question of justification arises concerning either giving up beliefs or adding to them. We may, at best, modify our habits of belief formation but there can be no issue concerning choice among beliefs. Sympathy with this attitude may account for the exaggerated emphasis on procedure, process and etiquette against which I just recently protested. Before pursuing the main line of discussion, a brief digression focusing on antivoluntarism and related issues may be in order.

We undoubtedly lack control over our doxastic and affective feelings as well as our dispositions to assent upon interrogation. If believing were a matter of passionate conviction as Hume and Berlin seem to say, expecting us to augment or diminish beliefs on demand would be absurd.

Beliefs, as I understand them here, are resources for deliberation. We use them as a standard for assessing propositions with respect to serious possibility. They enable us to discriminate the possibly true sheep from the certainly false goats. I fully believe that I am not a product of a virgin birth. In virtue of this conviction, I am *committed* to ruling out the logical possibility that I am the product of a virgin birth in any context of deliberation or inquiry that arises – as long as I retain the conviction. A full belief is to be understood, therefore, in terms of the doxastic commitments it generates and not in terms of the intensity of feelings accompanying such full belief. I grant that we lack control over the displays of doxastic feeling from which we suffer. More generally, I concede that our control of our dispositions to assent and their manifestation is often very indirect. And to the extent that acquiring dispositions and feelings of various sorts implements doxastic commitments and undertakings, we lack control over the fulfillment of the commitments we undertake. From this it does not follow that we lack control over our doxastic commitments.[1]

I neither know nor care whether my use of "belief" is ordinary or not. What matters is that those resources for deliberation we take for granted in the manner indicated are of great relevance to the evaluation of our conduct both in practical deliberation and in scientific inquiry as well; so it seems to me, in efforts to create and reconstruct works of art. By way of contrast, the liveliness and vivacity of our impressions, the firmness of our dispositions to answer questions when asked do not, in any obvious way, have much bearing on these matters. They may have an impact on the quality of our experience, but the quality of our experience is itself more likely to be enhanced if the aesthetics of experience is not made the center of attention. We should focus instead on the activities involved in inquiry and deliberation and the ends these activities are designed to promote.

Belief is a means or resource for deliberation. It is not a consummatory

1 The relation between commitment and performance in belief and in other attitudes is discussed in Levi, 1980 (ch. 1, sec. 5; ch. 8, sec. 1; and ch. 9, sec. 2) and in Levi, 1991 (ch. 2).

experience. *Qua* means it is fair game for deliberation addressed to its improvement. Perhaps, we may also enjoy our beliefs. I do not want to be a doxastic killjoy by ruling out that possibility. What I deplore is the epistemological aestheticism implicit in the Hume-Quine view that we lack control over our beliefs because our beliefs are feelings or dispositions beyond our control to modify although open to us to appreciate.

Returning from the digression to the main line of argument, given that we are speaking of beliefs in a sense in which beliefs are subject to control, my contention is that justification of ceasing to believe is as problematic as justification of coming to believe.

Berlin seems to take a different view. He claims that to open one's mind to the views of others and to take them seriously is always preferable to maintaining a closed mind and merely tolerating disagreement.

According to this position, we ought ideally to contract our beliefs to those uncontroverted and noncontroversial assumptions we share with our fellow citizens and, perhaps, the entire human race. If we managed to do this, the need for toleration would evaporate; for we would all be like-minded. Disagreement would collapse into empty mindedness. There are very few noncontroversial assumptions. Is not the threat of such empty mindedness the very abyss that worried Mill?

Even these few shared convictions are subject to the corrosive acid of scepticism. They may not be disputed at present, but they might be controverted in the future or they may have been disputed in the past. If we take a sufficiently catholic view of the community of beings to be respected, firm beliefs would be restricted to those which no one has, will or could deny.

I do not care to distinguish between the many variants of such sceptical advocacy of the empty mind as the ideal of rationality. In passing, I shall focus on a special case.

Some philosophers think we can avoid the abyss of empty-mindedness by denying that equating serious possibility with logical possibility is tantamount to such empty-mindedness. They see salvation in distinguishing possibilities with respect to probability.

I fail to see how such views help. Their advocates seem to think that it is dogmatic to be certain that one is not the product of a virgin birth but it is not dogmatic to assign a probability $1-\epsilon$ for small ϵ to that hypothesis. But suppose the harbinger of glad tidings comes along and declares with probability $1-\epsilon$ that I am the fruit of a virgin birth after all. Perhaps, he assigns probability 0.5. Is my disagreement with him any less when I assign probability $1-\epsilon$ to the proposition that I am not the son of a virgin than it would be if I assigned that proposition probability 1?

To be sure, disputes over appraisals of propositions with respect to probability are not disputes over the truth values of truth value bearing propositions. They remain, nonetheless, disagreements in probability judgment that have relevance to conduct. If I dismiss the messenger's glad tidings out of hand, I am showing disrespect for his views just as surely as I would if I were to dismiss the hypothesis to which he assigns probability of 0.5 as not a serious possibility.

It may be suggested that in the face of such disagreement, I should move to a state of suspense between my initial probability judgment and the judgment of the dissenter. Having done this, I should then assign probabilities to the two alternatives.

Thus, if the harbinger of glad tidings assigns probability $1-\epsilon$ to the hypothesis of my virgin birth and I assign probability ϵ to the same hypothesis, I might assign probability α to his probability judgment and $1-\alpha$ to my own. In that case, the probability I should assign to my virgin birth would be $\alpha(1-\epsilon)+(1-\alpha)\epsilon$. As long as α is positive but less than 1, this sum differs from both $1-\epsilon$ and ϵ. So assigning probabilities to the different probability judgments contradicts the assumption that one or the other of the competing probability judgments is the "right" one. But it is only on the basis of such an assumption of suspense between rival probability judgments that it makes sense to assign probabilities to the probabilities.

Such suspense is incoherent. One might, however, open one's mind to the probability judgments of others by shifting to indeterminate probability judgments. What this means and the extent of the indeterminacy is discussed in Levi, 1980.

Technical details need not concern us here. It is, nonetheless, important to recognize that the absurdity of giving a hearing to every rival view is not mitigated by urging everyone to regard every logical possibility to be seriously possible with varying degrees of probability. To assign definite probabilities is to be just as opinionated as it is to be certain. And to open up one's mind by going indeterminate in probability judgment leads to empty mindedness just as surely as ceasing to regard a proposition as certain does when one opens up one's mind promiscuously.

Much the same may be said when we turn to evaluations of propositions as better or worse relative to our moral commitments, political stances, economic goals and aesthetic attitudes. Such evaluations lack truth values just like probability judgments. But giving a hearing to every evaluation rival to one's own is just as absurd as is giving a hearing to every rival probability judgment or appraisal of possibility. Indiscriminate open mindedness is tantamount to empty mindedness in art, politics and morals as it is in science. I shall continue to speak of what is to be believed to be true or taken for granted and not about other forms of appraisal relevant to human concerns. But these remarks are intended to apply quite generally to other forms of appraisal as well.

We should not open up our minds on demand. Opening up one's mind incurs a loss of information which is settled and beyond serious doubt. Some inducement should be offered to compensate for such loss before incurring it can be justified. In this sense, we ought to require good reasons for giving a hearing to the dissenting views of others. Such good reasons should justify our incurring the losses of information entailed by doing so.

What inducement can justify giving up an assumption taken to be certainly true if mere disagreement with others is insufficient?

Even though no one ought to be so empty minded as to suspend judgment on every issue, no one should claim to have an answer to every question. We inquire because we are ignorant about some things even though we do claim to know others.

Sometimes we can justify adding new information to our knowledge on the basis of what we already know. Adopting a potential answer yields new information while risking error. Potential answers to a question are evaluated in terms of the extent to which the information obtained compensates for the risk incurred. Adopting a potential answer is justified if the compensation for risk of error afforded by the information promised is the best one available.

Such inductive inference is not the only way in which new information is added to knowledge or settled belief. A more common procedure is to rely on the testimony of one's own senses or on the testimony of expert witnesses. Consulting oracles in one of these ways is defensible provided we assume in advance that the program we adopt for processing the testimony of the senses or witnesses is sufficiently reliable. Unlike inferential expansion, such routine expansion can lead to the importation of information inconsistent with the background of settled assumptions into that background.[2] Experimental results and expert witnesses sometimes confound expectations which form part of the background against which we consult them.

Such conflict could be avoided by refusing to consult our senses or the testimony of witnesses. The cost of such refusal would be considerable.

2 For further discussion of the distinction between inferential and deliberate expansion, see Levi, 1980 (ch. 2), and Levi, 1991 (ch. 3).

We would deprive ourselves of information required to settle questions of interest that cannot be settled by deliberate or inductive inference alone.

Scientific practice refuses to dispense with the testimony of the senses. It seeks instead to restrict its trust to procedures of known high reliability and to improve techniques of observation and experimentation as much as is feasible. Such practice acknowledges that the testimony of the senses is a poor foundation for knowledge – counter to empiricist doctrine. We put up with the testimony of the senses as a source of information (suitably corrected by those adjustments for systematic error we can muster) so as to obtain information we cannot acquire easily in other ways.

The question of toleration vs. respect for dissenting opinion addressed by Berlin does not, of course, focus on the dissent of the senses. Dissent of witnesses is more germane. But the testimony of witnesses is no more and no less a source of information in scientific inquiry than is the testimony of the senses. Considering the parallels between the two should instruct us on how to approach the question of toleration vs. respect.

Reliance on the testimony of witnesses is a necessary evil with which we put up to obtain worthwhile information. To be sure, we should seek to restrict our trust to reliable and competent witnesses; but if we put up with the infirmities of the testimony of others in order to obtain more knowledge, we must accept the possibility that sometimes such testimony will conflict with the views we initially held.

Disagreement with the testimony of a witness when one was committed in advance of the testimony to adding it to one's stock of full beliefs constitutes one kind of occasion where one can justify moving to a position of suspense between one's initially held views and the conflicting testimony of the witness. If a trusted expert or witness disagrees with views I already take for granted, I am faced with the problem of ridding my views of inconsistency. I must either modify my initially held convictions, question the testimony of the authority or witness and conceivably my judgment that the witness is usually reliable or, as will often prove best, question both the background information and the testimony of the witness.

In such cases, opening my mind by removing erstwhile held views to give a hearing to the dissenting opinion of another is justified by the respect I initially had for the competence of the witness or expert on the matter under consideration. I am not obliged to accord everyone such respect on every issue. Nor, indeed, is this respect to be accorded on any other basis than a considered judgment that the witness is sufficiently reliable to warrant using his or her testimony in a program for adding information to my stock of evidence available as a resource for subsequent inquiry.

Respect for the agent's qualifications does not oblige me to defer slavishly to his views when we disagree but only to open up my mind by shifting to a position of suspense and initiating inquiry to settle the controversy. Respect for the person's competence engenders sceptical respect for his or her view - not deference.

Other grounds sometimes warrant opening up one's mind to rival views. Suppose someone proposes a hypothesis challenging some settled conviction of mine. From my point of view, the hypothesis is certainly and infallibly false. Yet, it may have its attractions. Were it true, it would supply a coherent and systematic account of phenomena in some domain which my current doctrine fails to handle. Were I to act on the basis of wishful thinking, I would exchange current doctrine for this hypothesis in order to gratify my desire for such understanding.

Doing so, however, ignores the claims of truth by deliberately replacing what is known for sure to be true by what is known for sure to be in error. Instead of doing that, one may open one's mind by moving to a position where both the initial doctrine and the new one are taken to be serious possibilities. In this way, deliberately importing error into one's doctrine may be avoided. To be sure, I would be surrendering information of which I am certain (relative to my initial view). Losing such information is not, however, deliberately importing error. And the loss of information may be outweighed by the benefit of giving an informationally attractive hypothesis a hearing.[3]

Whether such a modification of knowledge is justified depends on how attractive the new hypothesis is and how much useful information is lost in giving up the old. There is a trade-off to be made, and the terms of the trade-off in each situation control the propriety of opening up one's mind.

This kind of reason for giving up an erstwhile settled assumption differs from the one considered previously. The justification for opening up one's mind does not derive from respect for the authority of the agent who proposed the new hypothesis. A scientist lacking respect for

3 It should, perhaps, be said that, according to my usage, information may be false as well as true. The informational value or attractiveness of a hypothesis reflects those features of the hypothesis other than its truth value or probability of truth that are considered of sufficient cognitive importance to justify risking error. Specificity, explanatory power and simplicity are, in this sense, informational values.

Velikovsky as an authority on anything could, nonetheless, recognize one of his many suggestions to be an interesting hypothesis meriting a hearing. In such a case, interest in the hypothesis generates respect for its proponent rather than the other way around.

In sum, there are two types of reasons which may justify opening up one's mind to the views of a dissenter.[4]

The first derives from respect for the competence and authority of the dissenter concerning the subject matter which is the locus of the dispute. In such cases, our respect for the authority of the dissenter induces us to give a hearing to his or her dissenting view and renders it worthy of a hearing.

The second reason derives not from respect for the competence of the dissenter but rather from recognition of the informational value of the view he or she has put forward. In such cases, respect for the dissenter's view generates respect for the dissenter.

Armed with these considerations, let us return to Berlin's view that sceptical respect for the views of others is preferable to contemptuous toleration which, in turn, is better than suppression of dissent.

I doubt whether one can always justify contemptuous toleration over suppression. When religious fundamentalists or totalitarian parties threaten to take over the instruments of power, there may be good reason to prevent the effort even to the extent of suppressing their views. But such suppression is not to be taken lightly. There are serious considerations affording a strong presumption against suppression of dissent.

No single individual or institution can rely on his, her or its own resources exclusively for obtaining information either via observation and experimentation or from the currently available lore on all matters that are of concern. We must rely on others (i.e., other persons, institutions and machines) for information and it is obviously desirable that they be competent and reliable. Hence a considerable part of the resources of society ought to be invested in the training of competent and reliable sources of information.

More to the point, competent authorities are to be encouraged to express their views on matters within their competence with minimal legal and social constraint - at least on occasions where the information they have to convey is of relevance to the issues under consideration.

In some societies, efforts are made to restrict the freedom of expression of competent authorities to matters concerning which they are

4 See Levi, 1980 (ch. 3), and Levi, 1991 (ch. 4) for further discussion of reasons for contraction.

competent. If this could be done without great cost to the adequate functioning of such experts or without moral injury, we could at least entertain the propriety of such constraints on free expression.

Advocates of freedom of expression may reasonably doubt whether a high quality cadre of competent experts can be effectively and efficiently maintained for long periods of time if each expert is confined to his own expertise or whether, to the extent that it is feasible, it can be done without grave moral costs.

This argument obviously deviates from the Millian line that freedom of expression ought to be grounded on a recognition of human fallibility. Such an appeal to fallibility coheres poorly with the epistemological stance I have been taking in this essay. Millians may object that the argument I have advanced favors restricting freedom to an aristocracy of merit in the sense that only experts on some subject or other are to be granted free speech.

I deny the charge. There is, I believe, a positive expectation that the overwhelming majority of healthy, mature human beings are capable of serving as reliable witnesses on some matters even when not extensively trained and, moreover, that it is desirable that many of these be trained in more distinctive skills. The gratuitous and widespread suppression of free expression may inhibit the use of ordinary human beings in those ordinary capacities in which they prove competent or the training of some of them to higher grades of competence.

These considerations afford a positive expectation, nothing more, in favor of contemptuous toleration as compared to suppression of dissent. They do so without favoring an aristocracy of merit.

Consideration of merit looms larger when the relation between contemptuous toleration and sceptical respect is considered.

The views of the competent ought to be taken seriously - at least concerning matters about which they are competent. We may indulge them in their desires for free expression on issues concerning which they lack authority; but we have no obligation to take them seriously except when they have earned our respect.

This is no unqualified endorsement of meritocracy. Sometimes proposals are made which, regardless of their provenance, deserve a hearing because they offer entertainable solutions to unsolved problems. They may be incompatible with received doctrine, but their merit warrants our opening up our minds to the possibility that they are true.

Yet, this qualification does not support the universal superiority of opening up one's mind over refusal to do so. Contemptuous toleration is sometimes preferable to sceptical respect. There is no warrant for my taking seriously the possibility of my virgin birth, and although it would be wrong to impose a legal ban against the publication of creationist views, there may not be good enough reasons to take them seriously or to encourage their dissemination.

Even if it is conceded that good reasons are required for opening up one's mind, it may still be objected that unless dissent is promoted and not merely tolerated, conjectures that are candidates for serious consideration will not be forthcoming. We may not be obliged to open up our minds to dissent without good reason, but we are obliged not merely to tolerate but to encourage and support the dissemination of dissenting views anyway.

In some situations, this line of argument makes good sense. Sometimes the inadequacy of current doctrine to gratify our demands for information is clear; there is little promise that it will do so in the future, and there is some urgency to having the demands for information gratified. In such cases, casting about for some rival to current orthodoxy that will do better is a sensible strategy. Keynesian theory was dismissed as inadequate by many for the purpose of prediction and control of economic development; for this reason, it became arguable that the dissemination and proliferation of a diversity of anti-Keynesian views was desirable.

I do not mean to endorse this verdict about Keynesianism. My point is that the positive desirability of proliferating new dissenting ideas has to be justified in each case. There is no automatic presumption that dissent is desirable or that it is better to have good reasons for opening one's mind to dissent than to have good reasons for tolerant contempt.

Paul Feyerabend (1978) disagrees. He contends that all current orthodoxies are inadequate to the demands being made upon them. Proliferation of eccentricity and opposition to orthodoxy is, for that reason alone, a strong desideratum in every case.

Feyerabend's charge depends for its cogency on his wholehearted commitment to a holistic outlook according to which one cannot endorse or reject one ingredient of a theory without significant reverberations throughout one's entire outlook. If a deficiency can be found anywhere in one's doctrine which might render the fostering of dissenting views desirable, dissent should be promoted concerning other aspects of the view as well. Since no aspect of the current stage of the evolving doctrine is adequate to its purpose, we should promote as much dissent as possible and assess the merits of opening our minds to it.

The prospect of a society organized in this fashion is hard to envisage in detail, but the gross impact is clear enough. Imagine the N.S.F. so reorganized as to consign a fraction of its funds to Velikovskian studies,

Dianetics, holistic medicine, creationism and other outlooks Feyerabend would encourage us to consider in a serious and sustained manner. The intellectual Babel resulting should provoke both hilarity and horror. Such Dadaism is more likely to promote confused minds than open minds. Worse, in the confusion, opportunities for suppression of dissent may lurk more ominously than the mere refusal to encourage dissent.

Popperian falsificationism may also be invoked in support of Feyerabend's advocacy of proliferation. Even if we cannot claim all current doctrine to be in trouble, we should acknowledge the possibility that it is so and, hence, promote in all ways feasible efforts to undermine such doctrine.

Mill argued that suppression of dissent presupposes the infallibility of one's views. Feyerabend goes further. He claims that the assumption of the fallibility of knowledge argues for the need for proliferation. I have already suggested that the presumption of infallibilism has little relevance to the issue of censorship. But the implications of Feyerabendian proliferation afford another strong argument as to why an agent from his or her point of view should regard all items of settled doctrine as infallibly true - i.e., as not possibly false in any sense seriously relevant to his or her conduct or inquiries.

In my judgment, there is little to choose between Feyerabend's saucy promotion of dadaistic proliferation and Berlin's urbane defense of sceptical respect. Both admire Mill and do so not merely because of his willingness to tolerate dissent but because they see him as seeking social arrangements that encourage it. Berlin is concerned to avoid the abyss of the empty mind but seems to think it can be done by conflating passionate intensity of feeling with deep conviction. Feyerabend equates the empty mind with the free spirit and so celebrates the abyss. The implications do not strike me as very attractive. Freedom from interference with the expression of heterodox views ought to be a fixed point of our political and social arrangements. There are no categorical freedoms in my view, but this one ought to be a reasonable facsimile thereof.

Nonetheless, we should stand opposed to the blanket presumption that sceptical respect is preferable to contemptuous toleration, and we should be just as adamant in rejecting the claim that where such respect is not preferable, conditions should be altered so as to make it so.

To be sure, we should foster the development of citizens who have a competent control of their intelligence meriting our respect. Such a citizenry is a benefit to everyone. But the indiscriminate promotion of dissent and its dissemination, whether in science, in morals, in politics or in art is a caricature of liberty encouraged by epistemologies grounded in misguided fallibilisms or fantastic holisms whose incoherencies are amply documented in the contemporary literature.

REFERENCES

Berlin, I. (1953), *The Hedgehog and the Fox,* Simon and Schuster.
(1969), *Four Essays on Liberty,* Oxford University Press.
Feyerabend, P. (1978), *Against Method,* Verso.
Levi, I. (1980), *The Enterprise of Knowledge,* MIT Press.
(1991), *The Fixation of Belief and Its Undoing,* Cambridge University Press.

Name Index

Subject Index